Aging and Work

Essential Reading from Taylor & Francis:

Working with Age
Edited by J Marquie, University of Tolouse Le Marie, France,
D. Paumes Cau-Bareille, INRETS, France and S. Volkoff, CREAPT, France
Hbk 0-415-24641-5

Work & Aging
Jan Snel, University of Amsterdam, The Netherlands & Roel
Cremer, Institution for Work Integration & Training, The Netherlands
Hb 0-7484-0164-4
Pb 0-7484-0165-2

Information and ordering details
For price availability and ordering visit our website
www.ergonomicsarena.com
Alternatively our books are available from all good bookshops

Aging and Work

Editor in Chief
Masaharu Kumashiro

Board of Editors
Tom Cox
Willem Goedhard
Juhani Ilmarinen

CRC Press
Taylor & Francis Group
Boca Raton London New York

CRC Press is an imprint of the
Taylor & Francis Group, an **informa** business

A TAYLOR & FRANCIS BOOK

First published 2003 by Taylor & Francis

Published 2019 by CRC Press
Taylor & Francis Group
6000 Broken Sound Parkway NW, Suite 300
Boca Raton, FL 33487-2742

© 2003 by Taylor & Francis Group, LLC
CRC Press is an imprint of Taylor & Francis Group, an Informa business

First issued in paperback 2019

No claim to original U.S. Government works

ISBN-13: 978-0-367-45469-2 (pbk)
ISBN-13: 978-0-415-27478-4 (hbk)

Visit the Taylor & Francis Web site at
http://www.taylorandfrancis.com

and the CRC Press Web site at
http://www.crcpress.com

This book has been produced from camera ready copy supplied by the authors

Every effort has been made to ensure that the advice and information in this book is true and accurate at the time of going to press. However, neither the publisher nor the authors can accept any legal responsibility or liability for any errors or omissions that may be made. In the case of drug administration, any medical procedure or the use of technical equipment mentioned within this book, you are strongly advised to consult the manufacturer's guidelines.

British Library Cataloguing in Publication Data
A catalogue record for this book is available from the British Library

Library of Congress Cataloging in Publication Data
Aging & work / editor in chief, Masaharu Kumashiro.
  p. cm.
  "A collection of selected papers from an international conference on aging and work that was held on September 26th–28th, 2001. This conference was incorporated within the 21st UOEH and the 4th IIES International Symposium 'Occupational health for the 21st century: new development in new directions'" – Pref.
  Includes bibliographical references and indexes.
  1. Aged – Employment – Congresses. 2. Age and employment – Congresses. 3. Aged – Employment – Government policy – Case studies – Congresses. I. Title: Aging and work. II. Kumashiro, Masaharu, 1947–

HD6279 .A335 2003
331.3'98–dc21   2002072013

Contents

Preface ix

Acknowledgments xiii

PART I   General Issues and Governmental Policy

1. Japanese Initiatives on Aging and Work: An Occupational Ergonomics
   Approach to Solving this Complex Problem 1
   Masaharu Kumashiro

2. Occupational Gerontology: the Science Aimed at Older Employees 9
   Willem J.A. Goedhard

3. Promotion of Work Ability during Aging 21
   Juhani Ilmarinen

4. Terminology of Aging Used in Legislation and Governmental Policy 37
   Seichi Horie, Takao Tsutsui

5. Employment of the Elderly in Korea 49
   Kwan S. Lee and Jae H. Kim

PART II   Measures for Healthy Aging: Lifestyle and Exercise

6. Lifespan Functional Fitness: Encouraging Human Struggle
   (Physical Activity) and Warning About the Cost of Technology 55
   Max Vercruyssen

7. Health Status and Lifestyles of Elderly Japanese Workers 65
   *Takashi Muto, Hidehiro Sugisawa, Hye-kyung Kim, Erika Kobayashi,
   Taro Fukaya, Yoko Sugihara, Hiroshi Shibata*

8. A Continuous Exercise Time and Psych-Physiological Reaction for a
   Suitable Prescriptive Exercise Program 77
   *Akiko Yamashita, Mitsuyuki Kawakami, Mari Watanabe,
   Yasumitsu Toba*

9. The Effect of Aging on Lipid Metabolism and Aerobic Ability 89
   *Yasumitsu Toba, Akiko Yamashita, Mitsuyuki Kawakami,
   Shigenobu Arai, Shizuo Sakamoto, Toshihiko Iijima*

10. For Whom is a Disability Pension a Good Solution When
    Musculoskeletal Disorders Prevent Work? 99
    Lena Edén, Göran Ejlertsson, Jan Petersson

PART III   Age-conscious Personnel Policies and Productive Aging

11. The Role of the Psychosocial Environment in Promoting the Health
and Performance of Older Workers 111
Amanda Griffiths

12. The Management of Work-related Stress with Regards to the Health
of Older Workers 119
Tom Cox

13. Post-Polio Fatigue and Aging: A New Problem in the Workplace
in Japan 129
*Satoru Saeki, Jin Takemura, Keinosuke Aridome, Yasuyuki Matsushima,
Hiromi Chisaka, Kenji Hachisuka*

14. Work Situation Evaluation as a Prerequisite for Productive Aging
of Engineers and Innovators 137
Klaus-Dieter Fröhner

15. Intergenerational Relations at Work in Sweden and the UK 143
Ingrid Johansson

16. A Work - Family Balance Approach to Research on Late Career
Workers 153
Martin M. Greller, Linda K. Stroh

17. Work Climate and the Age-Hostile Workplace 163
*Jacqueline Agnew, Gilbert C. Gee, David J. Laflamme,
Karen A. McDonnell, Barbara A. Curbow*

18. Occupational Activity and Aging 171
Sheng Wang

PART IV   Maintaining Work Ability of Elderly Workers

19. Company-Level Strategies for Promotion of Well-Being,
Work Ability and Total Productivity 177
Ove Näsman

20. Changes in the Work Ability Index of Aging Workers Related
to Participation in Activities for Promoting Health and Work Ability:
A 3-Year Program 185
Veikko Louhevaara, Anneli Leppänen, Soili Klemola

21. The Strict Agricultural Products Standard and the Difficulty
of Agricultural Work for Aged Workers 193
Yoshie Shimodaira, Nobuo Ohashi

22. Survey of Prospects for Elderly Care Workers 205
Hisao Nagata, Sunyoung Lee

23. A Program to Support and Maintain the Work Ability and Well-being
   of Kitchen Workers 213
   Leila Hopsu, Anneli Leppänen, Soili Klemola

PART V   Support Systems for Elderly Workers

24. Developing a New Work System for Aging Workers 223
   Mitsuyuki Kawakami

25. The Theory and Practice of Work Re-design in Small and
   Medium-sized Manufacturing Enterprises in an Aging Society 233
   Koki Mikami

26. The Anthropometric Data of Aging Workers in Taiwan 245
   Mao-Jiun J. Wang, Eric Ming-Yang Wang, and Yu-Cheng Lin

27. Development of a Work Support Tool for the Old with Work
   Postures as an Index 253
   *Masahiro Shibuya, Koki Mikami, Mitsuyuki Kawakami,
   Masaharu Kumashiro*

28. A Study on the Usability of Mobile Phones for the Elderly 261
   Kwan Suk Lee, Bohyun Kim

29. Ergonomics Problems in Job Redesign for Small-to-Medium
   Sized Factories 271
   Tetsuya Hasegawa, Masaharu Kumashiro

PART VI   Occupational Accidents and Incidents

30. A Study of Work Accidents in Fishery Work 281
   *Shuji Hisamune, Kiyoshi Amagai, Nobuo Kimura, Junji Kawasaki,
   Koya Kishida*

31. Renal Function Decline in Aged Workers Enhances Toxic Effect
   of Occupational Chemicals 291
   *Kan Usuda, Koichi Kono, Takemasa Watanabe, Tomotaro Dote, Hiroyasu
   Shimizu, Chisato Koizumi, Mika Tominaga, Mitsuya Akashi*

32. Characteristics and Perspectives of Occupational Accidents
   Involving Aged Workers in Korea 301
   Hyeon-Kyo Lim, Masaharu Kumashiro

Author Index 311

Keyword Index 313

Preface

This volume is a collection of selected papers from an international Conference on aging and work that was held on September 26th - 28th, 2001. This Conference was incorporated within The 21st UOEH and the 4th IIES International Symposium "Occupational Health for the 21st Century: New development in New Directions." Thus, issues relating to aging and work were taken up as one of the important strategies for Occupational Health activities for the 21st century. The Conference was co-sponsored by the 3rd International ICOH Conference on Aging and Work, Conference of the IEA Technical Committee for Safety and Health, Conference of the IEA Technical Committee for Aging, and The Association of Employment Development for Senior Citizens.

Although there is particular concern with issues relating to aging and work in well-developed regions, it is not often, within the broad field of gerontology, that interest is shown in elderly workers/aging in the workplace. To illustrate this, more than 3,500 people from 80 countries participated in The 17th Congress of the International Association of Gerontology held in Vancouver in 2001, and countless papers were presented. However, there were only nine papers (one session four papers and one invited Symposium six papers (of which one paper was canceled) presented on the subject of elderly workers/aging in the workplace. At the present time, it appears that efforts relating to elderly workers/aging in the workplace depend largely on the fields of ergonomics and occupational health. Today, as cries are raised about the enormous effects of the issue of aging and work on the labor market, society and the economy, the ergonomics and occupational health fields, which are tackling these issues head-on, are faced with big expectations and a heavy responsibility. In this context, it is hoped that Occupational Gerontology, advocated in this book by Dr. Willem Goedhard, will be established. Further, there is no doubt that the key words in this research, "work ability" and "employability," will play important roles.

This volume introduces representative administrative activities in Japan, where the population is aging at the fastest rate in the world. It also introduces the current status of research on aging and work measures in the East Asian countries of Korea, Taiwan and China. In addition, the concept of PWA (Promotion of Work Ability) is introduced. The PWA concept, which combines researchers in the three fields of occupational health, ergonomics, and aging and work, has been promoted and studied for a number of years by the government of Finland, one of the most advanced countries in the area of aging and work measures. Many international cases relating to the WAI (Work Ability Index), developed based on the PWA concept, are presented. It is worth noting that the WAI currently is translated into 14 languages and is becoming the global model for existing work ability diagnosis checklists.

Much of the research carried out with regard to aging in the workplace concentrates on activities from the Occupational Health field relating to employment of aging workers and health, but the approach from the psychological viewpoint also is extremely interesting. Here, we also look at psychosocial factors relating to work and organizations, and the relationships between age and work ability and health. Further, with regard to the health of aging workers, we look at physical exercise as protection against obesity and obesity-related illnesses as the quid pro quo of technological innovations.

The discussion of countermeasures at the corporate level are interspersed with concrete examples. In particular, there are amazing reports from Japan on the progress of "kaizen" (= improvement), which is the specialty of Japanese companies.

In this way, measures for aging and work on the individual, work place, corporate and national levels are included in this collection of papers on aging and work, which thus should satisfy a wide readership, from researchers and students to business people.

Finally, the Editor-in-Chief presents here the Occupational Health Charter that was proposed and adopted at the Conference.

This Charter, taking into account the health aspects of aging workers, relates to Occupational Ergonomics for the purpose of helping aging workers to make full use of their experience and abilities in moving toward a lively social economy. The subject age range of this Charter for aging workers is defined to include middle-aged workers and all those aged 45 or older. The upper end of the age range is 65 years, taking into account productive age and retirement age.

Strategy: To introduce the theory and practice of Occupational Ergonomics in labor economy policies that respond to the aging society, and to work toward the coexistence of humanization of labor and improved productivity.

TASK 1: *To facilitate working conditions that are easy on the minds and bodies of aging workers, and to make it easier to set up such working environments.*

We will evaluate the physical functioning levels (functional age in the workplace, etc.) directly connected to the skills of aging workers, based on an accurate grasp of the health standards that affect their skills. We will use these results to establish reasonable labor loads that effectively utilize the work ability of aging workers. Moreover, we will take steps to standardize guidelines so that the assessed work ability for one company can apply to various other firms and work environments in the external labor market, leading to further employability.

TASK 2: *To promote an organizational design and administration for building a mixed-age workplace in which both younger and older workers can enjoy pursuing productive activities with no thought of age.*

An organizational design will be developed for a mixed-age workplace where older and younger workers can do their jobs without concern about age differences. Toward this end, we must grasp objectively the flexibility that diminishes with age, and conduct research into and development of rational support systems (so called gerontechnology). By the same token, the organization must smoothly pass on to the younger workers synergetic skills, those interactive capabilities that stabilize and even improve with aging, and so we will create a system that facilitates communication between the younger and older groups.

TASK 3: To set up the conditions under which not only aging workers but every worker can easily customize his or her own duties, such as job description, working environment, and the like.

We will gather various examples of workplace improvements implemented by different companies into a single database to create systems whereby dissimilar industries and workplaces can progress together. The data will include not only improvement results and case studies but also learning on the techniques used as well as lists of advisors by field.

TASK 4: To build an education and training system enabling prompt responses to technological innovations.

To enable aging workers to better handle information technology (IT) tasks, we will create an education and training system that takes into account the characteristics of aging, with the aim to introduce this useful instruction at schools as well as in the workplace. We will promote health management techniques that deter mental problems such as techno-stress, the anxiety or fatigue that may develop from working in an IT environment. Moreover, we will endeavor to prevent the younger and the older workers on each side of the ever-widening information technology gap from losing touch with each other.

Masaharu Kumashiro, Ph.D.
Editor-in-Cheif

Acknowledgments

This is a compilation of the papers presented at the International Conference on Aging and Work held for the three days from Wednesday, September 26, to Friday, September 28, 2001. This conference for aging and work was held as part of the 21st UOEH and the 4th IIES International Symposium, "Occupational Health for the 21st Century: New Developments in New Directions". The conference was sponsored by the ICOH Scientific Committee for Aging and Work and the IEA Technical Committee for Safety and Health. We were able to hold the conference and publish these proceedings due to the substantial financial assistance we received not only from the University of Occupational and Environmental Health, Japan, but also from the Society for International Cooperation in Occupational Health, Kitakyushu, a volunteer organization comprised of companies in Kitakyushu for supporting occupational health activities, and the Association of Employment for Senior Citizens, an organization affiliated with the Ministry of Health, Labor, and Welfare.

In addition, Professor Willem J.A. Goedhard, Professor Juhani Ilmarinen, and Professor Tom Cox rendered invaluable assistance regarding the scientific aspects of the conference and in editing the proceedings. I would like to express my thanks to these organizations and people for their help.

Further, I want to express my thanks to Mr. Osami Hagiwara and Ms. Akiko Inage of the Alphacom Co., Ltd., which served as the secretariat for the conference and provided their assistance from the operation and administration of the conference to publishing the proceedings. Finally, I also want to thank Mr. Tony Moore and Ms. Sarah Kramer of Taylor & Francis for their understanding and support of the plans to publish these papers.

Masaharu Kumashiro, Ph.D.
May 2002

PART I
General Issues and Governmental Policy

1. Japanese Initiatives on Aging and Work: An Occupational Ergonomics Approach to Solving this Complex Problem

Masaharu Kumashiro

Department of Ergonomics, IIES,
University of Occupational and Environmental Health, Kitakyushu, Japan

ABSTRACT

The first half of the 21st century is said to be a time for dealing with population aging. It has been forecasted that the aging of populations of advanced nations will accelerate during the first several decades of the century, and that by the latter part this phenomenon will have become pronounced in developing nations. The advent of an aging society decreases the population of young workers in the labor market, thus placing greater pressure on the supply of labor by aging workers (45 years of age and above). As a result, the formulation of strategies on "productive aging" aimed at the effective use of aging workers has become an urgent task.

In particular, the aging of Japan's labor population is proceeding at a pace not seen anywhere else in the world. As measures to cope with this rapid aging of Japan's workforce, for examples, the Ministry of Labor proposed in its 1997 White Paper on Labor, three tasks to expedite the utilization of Japan's elderly work force: (1) consideration of health issues, (2) review of wage systems and retirement pay, (3) and re-evaluation of job content and the workplace environment. To satisfy those three conditions, tools for work ability assessment must be developed.

It is in this connection that the Japanese government, under the auspices of the Ministry of Health, Welfare and Labor, has established the Millennium Project - a project which is examining 12 issues seen as means of resolving the three tasks mentioned above (April 2000-March 2001). Meanwhile, the author has also developed a method for work ability diagnosis and evaluation as one step towards the Ergoma strategy and actions for achieving productive aging (Kumashiro, 1999). The essence of this method is the formula $P = f(W, A, C, M)$ where P is the value of labor performed by aging workers, with a substantial economic effect; W is the work conditions and environment (including a database of success stories); A is the capacity for duties (skill plus knowledge plus experience); C is the ability to adapt to work (one's functional age); and M is the worker's motivation (desire to perform work). This paper will first introduce an outline of measures adopted by the Japanese government related to aging workers. This will be followed by a discussion of the Ergoma strategy formulated by the author as a strategy aimed at resolving aging worker-related issues.

Keywords: *Work ability, Ergoma strategy, Productive aging, Human resources*

INTRODUCTION

The beginning of the 21st century has been called a time for implementing measures aimed at solving problems caused by population aging. It has been forecast that the aging of populations of advanced nations will accelerate during the first several decades of the century, and that by the latter part this phenomenon will have become pronounced in developing nations as well.

The advent of an aging society brings about a decline in the population of younger workers in the labor market, thus increasing pressure on the supply of labor by aging workers. This has given rise to a pressing need to establish strategies aimed at the effective utilization of the labor of aging workers. An objective examination and evaluation of the flexible capabilities and the synergetic skills, which become more pronounced with age, must be undertaken as a basis for creating work conditions and a work environment which enable the aging worker to maintain good health and to be able to work safely. The provision of an employment environment which allows the aging worker to work comfortably not only raises the productivity of enterprises, but is also a source for promoting national economic growth. But first of all, any discussion on the aging worker requires a clear definition of the age group into which these workers fall. This paper will focus on aging workers defined as those in the age range of 45 to 65 years of age.

This is an important point. The World Health Organization defines "aging at the workplace" and "elderly workers" as pertaining to persons 45 years of age and older (WHO, 1993). For example, the laws of Japan's Ministry of Health, Labor and Welfare relating to employment define "moderately aged" workers as being between 45 and 54, and aged workers as being 55 or older.

On the other hand, the working-age population is defined as persons between the ages of 15 and 64. Matters relating to "aging at the workplace" or "elderly workers" should be interpreted as applying to persons between the ages of 45 and 64. However, because the Japanese government is currently considering a retirement age of 65, the age range of between 45 and 65 has been adopted as the age range of elderly or aging workers for the purposes of this discussion.

WHY MUST ERGONOMISTS SHOW MORE INTEREST IN ISSUES RELATING TO AGING WORKERS?

In Japan, a country already possessing an aging workforce, there is a need for healthy companies, which endeavor to raise true productivity while striving to attain a harmony between the humanization of labor and productivity gains. The four fundamental elements of environment, safety, health, and efficiency must be present in order to become a healthy company. The first step to meeting this challenge is to achieve an objective understanding of human characteristics, especially those of aging, to use as a basis for formulating economic strategies that take adequate account of the advantages and disadvantages that aging brings. In other words, the involvement of ergonomists is a pre-condition. It is to be hoped that ergonomic research will be undertaken as a basis for such economic strategies. This research would include an examination of the capacity to adapt to work so that there is no mismatch between the work capacity of middle-aged and elderly workers and the degree of qualitative and quantitative burden in the work imposed on this group of workers, and an evaluation of employment capacity on the basis of that examination. In particular, it is imperative that enterprises, which have come into contact with aging workers, introduce ergonomics so that their basic stance on individuality (difference), an extremely common characteristic of aging, can be moderated.

PRINCIPAL ACTIVITIES OF ERGONOMICS WITH RESPECT TO AGING

In light of this approach to an aged society, an increasing number of ergonomists are also beginning to show interest in the problems of aging. From the vantage point of the ergonomist, subjects employed in research on aging may be broadly divided into two types.

The first group of research subjects is of persons 65 years of age or older, and their nursing care providers. The emphasis of such research is placed mainly on the development of supporting equipment, devices and tools used to provide nursing care for the aged. The aim of this field of gerontechnology is to marshal technology for the benefit of the elderly and their nursing care providers.

The second group of research subjects consists of aging workers, roughly 45 to 65 years of age. Various research projects in this area are being conducted by the International Commission on Occupational Health (ICOH) as well as by the IEA. These studies span a broad range, including health promotion, work ability, development of supporting equipment for the workplace, and work improvement.

JAPANESE GOVERNMENT INITIATIVES

A discussion of some of the initiatives being undertaken by the Japanese government in relation to the aging worker, in particular measures being considered by Japan's Ministry of Labor, is provided below.

In its 1997 'White Paper on Labor', the Ministry of Labor proposed three major tasks for utilizing Japan's elderly labor force. The first task is the consideration of health issues, followed by a review of wage systems and retirement pay. This second task is related to the fact that, in general, Japan's wage system is based on a seniority system. The combination of reduced labor efficiency caused by work force aging and simultaneous increases in wages have together dealt a major blow to the Japanese economy. The last task is reconsideration of job contents and the work place environment.

A further initiative adopted to resolve issues raised by an aging work force is the Millennium Project, which was commenced in the summer of 2000 and scheduled to continue through to the end of 2001.

As part of the Millennium Project the Ministry of Health and Labor has been consigned to undertake research on twelve topics, which fall into three broad areas of research. These areas are basic research aimed at achieving an age-free society, research on a management model for elderly employment in relation to policies for promoting the systematisation of extending the employment of workers aged in their early 60's, and research on the creation of a work place which takes the elderly into account. The twelve topics provide an insight into the issues being considered by the Japanese government in relation to its policies on an aging work force. These topics are:

(1) The acquisition and analysis of early examples of the utilization of the elderly from within Japan and overseas;
(2) Joint research undertaken by Japanese and foreign experts related to economic and sociological measures adopted for the promotion of the employment of the elderly;
(3) Research related to the construction of models in the welfare and lifestyle-related service sectors in which the elderly are employed;
(4) Research and surveys related to restrictions on work load, which take into account the health and safety of elderly workers;
(5) Research related to VDT work, which takes into account the health management of elderly workers;
(6) Research related to policies for promoting the systematization of extending the employment of workers aged in their early 60's;
(7) Research related to the establishment of systems for examining work ability, which have been adopted in response to advances in information technology;
(8) Research related to the establishment of databases on the careers of elderly white collar workers;
(9) Research on training and training methods aimed at the elderly;
(10) Research related to the establishment of models for utilising elderly workers in the manufacturing sector;
(11) Research related to the establishment of models for utilising elderly workers in the nursing care sector;
(12) Research related to the establishment of support systems for work improvement which make use of the Web.

APPROACHES TO HUMAN RESOURCES FROM THE PERSPECTIVE OF WORK ABILITY

In 1981 in Finland a Work Ability Index was developed as a method of evaluating the work ability of elderly workers (Ilmarinen, 1999). Today this Work Ability Index is used in twelve countries around the world. The approach to work ability manifest in the index is, at an international level, probably the most popular of any. Because according to the Work Ability Index an appropriate examination and evaluation of work ability is also an effective tool for the examination and evaluation of employability (employability which is also applicable to the external labor market), which is an important indicator of the promotion of employment, it is one of the main approaches which will be adopted for promoting future labor policies in Japan. As shown in Figure 1.1, the author has proposed an approach for the examination and evaluation of employability that differs from the Work Ability Index (Kumashiro, 2000). This strategic plan firstly examines approaches to human resources, which comprises its foundation, while bearing some similarity to the approach to the concept of employability contained in the Work Ability Index.

The author's approach to work ability is based on the concept of human resources as shown in Figure 1.2. It can be seen that it is not vastly different from several theories that have been put forward in the past. There is a need to consider work capacity and its direct relationship to work in terms of blue collar and white collar workers. Here, blue collar indicates field workers, while white collar includes managers, IT engineers, and the like. There are three factors to be considered for each of these two groups.

The establishment of an appropriate approach for the evaluation of work ability may be regarded as the foundation on which to construct productive aging. If work ability can be assessed as a company-level strategy, it then becomes possible to evaluate employability as a national-level strategy, this making possible the creation and expansion of employment in an aging society.

ERGOMA (ERGONOMICS AND INDUSTRIAL MANAGEMENT) STRATEGIES AND ACTIONS FOR ACHIEVING PRODUCTIVE AGING

The formula proposed for some time now by the author for obtaining productive aging, and which is based on these theories, is shown in the diagram below. Its content is not new, as it has been presented on previous occasions. Figure 1.1 is an amended version of a strategy constructed by the author on the basis of the aforementioned approach to human resources for achieving productive aging (Kumashiro, 1999).

The formula is $P = f(W, A, C, M)$, where P is the value of labor performed by aging workers, with a substantial economic effect; W is the work conditions and environment, including job redesign; A is the work capacity - that is, skill plus knowledge plus experience; C is the functional capacity - in other words, the functional age of the workers; and M is the worker's motivation - that is, the desire to perform work.

The strategy and accompanied action proposed here aim at the creation of an "ageless and sexless" work place. It is precisely in an aging society that an ergonomic strategy for the overall working population, which is not unduly conscious of aging workers, is essential.

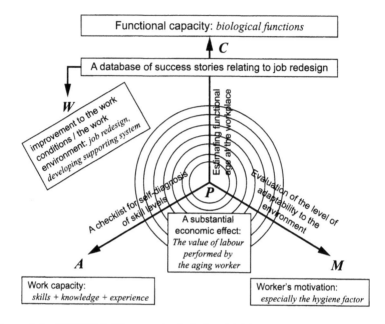

Figure 1.1   ERGOMA (Ergonomics in Industrial Management) strategy and actions for achieving productive aging (Kumashiro, 2000)

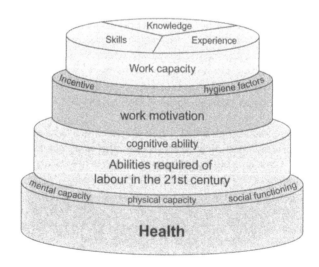

Figure 1.2-1 Human resources in the work place

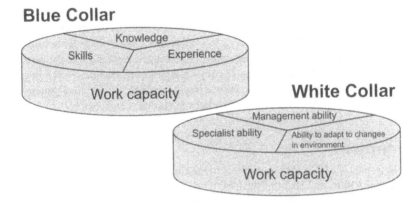

Figure 1.2-2 Human resources in the work place

CONCLUSION

The issues presented by an aging work force pose a major challenge, which must be tackled in order to avoid more serious regional crises. The challenge does not merely involve matters of occupational health, safety and ergonomics, which happen to concern both aging and work.

It is clear that the demographic shift toward old age in the populations of more developed regions will become a major encumbrance to the labor economy of the world. This in turn will have serious consequences for societies in the 21st century and the world economy. Unfortunately, only a few persons have warned that this state of affairs may contribute to the collapse of the world economy.

The author has adopted a set of strategies for coping with this situation, derived from the evaluation of the work capability and productivity of aging workers. This set of strategies is encapsulated in the equation presented earlier, $P = f(W, A, C, M)$. The capacity to execute duties is inferred to depend heavily on three factors: work capacity, functional capacity, and work motivation. In terms of internal factors (that is, characteristics of the individual), it has been assumed that the ability to execute duties, conceived as output, is determined by the interactions between these three factors.

Consequently, in planning the development of a system for the examination and evaluation of capacity to perform duties, diagnostic systems for each of these three factors, and for the evaluation of combinations thereof, are appropriate. A work capacity diagnosis system will primarily involve a quantitative estimation of an individual's accumulated skills, knowledge and experience relative to the jobs to be performed. Similarly, a system for the diagnosis of functional age will extract the various physiological functions most closely related to the work, and use these as indices for estimating functional age. And lastly, a system for the evaluation of work motivation employs a checklist to evaluate an individual's ability to cope with other workers in the organization, as well as the power to enhance his/her drive to work.

In conclusion, it is essential that ergonomists confront issues related to aging workers and an aging work force. It may even be that society places hopes on ergonomists, acting as decision makers, to resolve these issues. In order to live up to these expectations, the relevant issues must be examined from the vantage points of macroeconomics and a labor-based society, in addition to medical, psychological, and engineering perspectives. In this sense, it is quite probable that in future ergonomists may be obliged to address the interaction between humans and society.

REFERENCES

1. Ilmarinen, J., 1999, Aging Workers in the European Union, FIOH & Ministry of Social Affairs and Health, Helsinki.
2. Kumashiro, M., 1999, Strategy and Actions for Achieving Productive Aging in Japan, *Experimental Aging Research*, **25(4)**, pp.461-470.
3. Kumashiro, M., 2000, Ergonomics Strategies and Actions for Achieving Productive Use of an Aging Work Force, *Ergonomics 43(7)*, pp.1007-1018.
4. Ministry of Labor 1997, *Heisei 9-Nenenban Roudou Hakusho* (White Paper on Labor for 1997), Japan Institute of Labor, Tokyo, pp.30.
5. World Health Organization (WHO), 1993, Aging and Work Capacity, Report of a WHO Study Group, *WHO Technical Report Series 835*, World Health Organization, Geneva.

2. Occupational Gerontology: the Science Aimed at Older Employees

Willem J.A. Goedhard

The Netherlands Foundation of Occupational Health and Aging, Middelburg, The Netherlands

ABSTRACT

The ICOH scientific committee 'Aging and Work', established in 1989, has developed several initiatives and organised a series of conferences and workshops that resulted in the publication of many interesting research papers in scientific journals and congress proceedings. Important aims are the improvement of occupational health care of older workers, the maintenance of work ability and employability and the prevention of early retirement due to disability. These aims can only be achieved by thorough knowledge of (1) occupational health and (2) aging during the working life. This leads to the concept of Occupational Gerontology, situated between 'fundamental gerontology' and 'occupational health'. Occupational Gerontology will be aimed at the specific problems of aging workers (45 - 70+ years), as well as the development of new instruments and protocols. A good example is the Work Ability Index (WAI).

In this paper the new concept of the Working Life Expectancy (WLE) is being introduced. Based on average data of the Healthy Life Expectancy and data obtained from WAI assessments of individual workers, the theoretical WLE can be obtained. The WLE may be useful in determining the future risk of disability.

Keywords: *Work ability, Healthy life expectancy, Working life expectancy*

INTRODUCTION

The ICOH Scientific Committee 'Aging and Work' was established in 1989 under the auspices of the International Commission on Occupational Health. The most important objective was the development of optimum care for older employees. Although ICOH was founded in 1906, it was in existence for more than eight decades before the necessity for specific attention to older workers was felt. One can only speculate about the question of why aging was obviously not considered an important item of occupational health for quite a long period. The reason for this is not very clear at this moment, and may be subject to further research. Two aspects can be mentioned here as relevant: first, gerontology is a relatively young science [1], and second, older employees in industrialised countries are increasing in relative and absolute numbers due to the greying of the population [2].

GERONTOLOGY AND OCCUPATIONAL HEALTH

Regarding the specific problems of older employees in relation to their work environment, it is useful to realise that two different sciences are involved, namely gerontology and occupational health. Gerontology concentrates on the aging process of different species, including man. Important questions are, for example, the causes of deterioration of functions and organ systems with advancing age, and the questions of longevity and maximum life span. These questions pertain to fundamental gerontology, which can be defined as: 'The process of change in the organism, which decreases the capacity for adaptation to environmental demands'.

Aging is a very complex concept and a multifaceted problem. Not all observations can be brought into a single theory. There are several theories of aging; none of the present theories has been fully confirmed, but many are plausible and supported by experimental data [3]. There are two groups of theories:

1. Stochastic Theories (stochastic = arising from chance; aging = an accumulation of insults or damage caused by the environment)
2. Genetic (programmed, non-stochastic) Theories

Another classification is based on the assumption of Primary and Secondary Aging. Primary Aging can be defined as the result of inherited biological processes that are time-dependent. Secondary Aging is defined as a process that is caused by the decline in function because of unfavourable living and working conditions as well as diseases.

OCCUPATIONAL GERONTOLOGY: A NEW CONCEPT

Until recently, human gerontology has mainly been associated with the problems of the aged population, i.e., people past the age of retirement. Geriatric medicine emerged from the growing knowledge about age-associated changes of organ systems which eventually increase the risks of multiple pathology and diseases.

For actively working employees, occupational health has been recognised for more than a hundred years as an important science. In order to concentrate more specifically on relevant issues concerning older employees, it is useful to bridge the gap between occupational health and human gerontology. This can be achieved by the acknowledgment of a separate science that I propose to call Occupational Gerontology. Its position is visualised in Table 2.1. Sterns et al. have launched the concept of Industrial Gerontology, which may have similar implications [4].

Table 2.1   Position of Occupational Gerontology among Occupational Health and Fundamental Gerontology

Level of Interest	Research or Practical Activities
Occupational Health	1. Implementation of Prevention Programs for Aging Workers (e.g., fitness programs, secondary prevention programs) 2. Changes in the Work Environment of Aging Workers (e.g., ergonomic adaptation, exposure limits) 3. Changes in the Work Organisation of Aging Workers (training, gradual retirement programs)
Occupational Gerontology	1. Studies on the work ability of aging workers 2. Studies on the relationship between work environment and aging workers 3. Studies on the relationship between work organisation and aging workers
Fundamental Gerontology	1. Theories of aging 2. Studies on aging cells, tissues, organisms

Within occupational gerontology three branches of research can be distinguished, analogous to general gerontology: biological, psychological and social gerontology. These branches can then be studied in relation to the areas of occupational health that are important aspects for the maintenance of work ability [5]: the aging worker (health, life style, physical capacity), the work environment (physical work load) and the work organisation (psycho-social aspects, management, training and education) (see Table 2.2).

Table 2.2   Branches of Occupational Gerontology in relation to areas of interest in the maintenance of work ability of older employees

	Work Environment	Aging Worker	Work Organisation
Branches of Occupational Gerontology	Job content and exposure	Work Ability	Employability
(1) Biological Gerontology	- Ergonomics - Physical and Chemical Exposure	- Physical / Physiological Ability - Sensory-Perceptual Ability	- Shift work
(2) Psychological Gerontology	- Mental demands / exposures	- Psychomotor - Cognitive Ability	- Work stress
(3) Social Gerontology	- Social demands / exposures	- Social Ability	- Learning / Training - (Early) Retirement

WHAT IS THE DEFINITION OF AN OLDER EMPLOYEE?

During the 1990s the lower age limit of older employees was settled at 45 years [6]. The upper limit has not been settled. One might even wonder whether there is such an upper limit. For a long time, the age of 65 years has been the traditional age of retirement. During the last decades of the 20th century this age limit even decreased to 55-60 years when early retirement policies developed as a consequence of redundancies in many companies. This implied that the post-productivity period of many people increased considerably, which seems to be a kind of paradox with the increase in longevity. An example: in 1997 in the European Union (EU) the percentages of employed male workers of 55-59 years and 60-64 years were only 68% and 33% respectively [2]. Over the next few decades the average age of the work force in the EU will increase considerably. The same is observed in other industrialised countries, as well.

The length of the working life has shortened considerably during the 20th century, especially as a consequence of the longer education period for young people. At the other end of the working life, many workers are subject to a mandatory retirement age; also retirement pensions allow workers to leave the labour market with, until recently, relatively good prospects for a carefree third age.

At 65 years the average life expectancy is about 15 (men) - 18 years (women). It can be expected that in the long run many workers will have to remain employed beyond the age of 65 years because of shortages in manpower. It will then be a challenge for occupational health to preserve the health and the ability to work of these elderly employees.

The determinants of the working life are [7]:

1. Employment
2. Employability (employment policy, social & health services, prevention of age-discrimination)
3. Work Ability (human resources, work conditions).

WORK ABILITY OF OLDER EMPLOYEES

Work Ability is an important aspect in relation to older employees. The risk of a diminished level of work ability increases with age. In the Finnish Study 'Respect for the Ageing', work ability decreased in 30% of the workers over a study period of 11 years [8]. Occupational Health for older workers has benefited from the initiatives of scientists of the Finnish Institute of Occupational Health. They developed, in the early 1980s, an important new instrument called the Work Ability Index. This questionnaire was successfully tested in Finland and subsequently translated and applied in several other countries [9-12].

SOME EXPERIENCES WITH THE WORK ABILITY INDEX (WAI) IN THE NETHERLANDS

In 1995, the Work Ability Index (WAI) was translated into Dutch and a first pilot study was done among Dutch industrial and clerical workers. In 1998, the 2nd edition of the WAI was published by the Finnish Institute (FIOH) [13]. The following improvements had been implemented:

- 4 categories of Work Ability (instead of 3): poor, moderate, good, excellent (Table 2.3);
- Forms for periodic follow-up of the WAI score of individual workers became available;
- A list of follow-up measures was published; this list is an important tool for specific advice to individual workers and the management.?

This second version is now being implemented in the Netherlands and the Dutch-speaking part of Belgium under the auspices of the Netherlands Foundation of Occupational Health and Aging.

Table 2.3 Work ability categories by WAI score

WAI score	Work Ability
7-27	Poor
28-36	Moderate
37-43	Good
44-49	Excellent

The rate of aging: An important question concerns the rate of change in work ability in relation to age. Physiological functions (e.g., pulmonary function, maximal endurance capacity, maximal heart rate) usually decrease roughly by about 1% per year (30 years = 100%) [14]. It was found that the decrease in average WAI score with age is smaller than the average physiological decrease [10].

The WAI is a complex entity that may depend on many different variables, such as age, work loads and exposures, work organisation and individual health.

In this chapter some results of experiences with the WAI in the Netherlands are shown. An analysis is made of the changes in mean WAI scores of different age groups.

Furthermore, an analysis is made of the different items of the WAI. The WAI score is based on seven items, each with a different score.

METHODS

For the study, the WAI questionnaire was implemented in the Periodic Health Examinations of two Dutch Occupational Health services. The examined persons were industrial and clerical workers (age-range: 19-60 years; N=283).

RESULTS

The means and standard deviations (s.d.) of the WAI and its different items are shown in Table 2.4. The figures represent the study group as well as two sub-groups: (a) with a poor-moderate W1AI score and (b) with a good-excellent WAI score.

Table 2.4   Scores (mean ± s.d.) of the WAI and its different items: the study group (all) and two sub-groups 'poor-moderate' and 'good-excellent' respectively

Item	Range	score (all)	score (category poor-moderate)	score (category good-excellent)
1. (Present work ability compared with lifetime best)	0-10	8.7 ± 1.6	6.2 ± 1.9	9.1 ± 1.1
2. (Work ability in relation to physical and mental work demands)	2-10	8.4 ± 1.0	7.4 ± 1.1	8.6 ± 0.9
3. (Diseases)	1-7	5.1 ± 1.8	2.6 ± 1.4	5.5 ± 1.5
4. (Work impairment due to disease)	1-6	5.5 ± 1.0	3.8 ± 1.5	5.6 ± 0.5
5. (Sickness absence)	1-5	4.0 ± 1.0	2.9 ± 1.2	4.2 ± 0.8
6. (Prognosis of work ability)	1-7	6.9 ± 0.6	6.4 ± 1.2	7.0 ± 0.4
7. (Psychological resources)	1-4	3.6 ± 0.6	3.1 ± 0.8	3.6 ± 0.5
WAI score (all categories)	7-49	42.1 ± 5.1		
WAI score (poor-moderate)	7-36		32.2 ± 3.7	
WAI score (good-excellent)	37-49			43.7 ± 3.1
Age (years)	19-60	40.4 ± 9.6		
Age (years)	26-60		46.0 ± 8.6	
Age (years)	19-59			39.5 ± 9.5
Number		283	40	243

WORK ABILITY AND AGE

In Table 2.5 it is clearly shown that the average work ability of workers of different age groups decreases with age. In the youngest age group only 2% of the workers have a poor - moderate work ability; in the oldest age group 28% of the workers have a poor - moderate work ability.

Table 2.5 Percentages of WAI scores by age-group

Age-group (years)	20-29	30-39	40-49	50-59
WAI category				
7-27 (poor)	—	1%	3%	2%
28-36 (moderate)	2%	7%	14%	26%
37-43 (good)	34%	41%	40%	34%
44-49 (excellent)	64%	50%	44%	38%

WORKING LIFE EXPECTANCY: A NEW CONCEPT

An important issue in gerontology is the study of longevity of different species. In human gerontology three different parameters of longevity are distinguished:

1. Maximum life-span1.
   - 'The days of man will be 120 years.' (Genesis 6:3)
2. Mean life expectancy
   - at birth
   - at different ages
3. Healthy life expectancy
   - at birth
   - at different ages

ad (1): The studies on the maximum life span are not very relevant for this paper.

ad (2): Mean life expectancy has increased in many countries mainly due to a decrease in infant mortality, improvement in living and housing conditions (water supply, sewer systems). Table 2.6 shows the improvements in mean life span since ancient times (data derived from [15,16]).

Table 2.6 Mean life span (life expectancy) since ancient times

Population	Period	Mean life span (estimated)
Neanderthal	Pre-historic	29 years
Bronze age	3500 - 1000 BC	38 years
Classic Greece / Romans	300 BC-300 AD	35 years
England (after Bblack Ddeath)	1500	38 years
The Netherlands	1840	36(m) - 38.5(f) years
The Netherlands	1900	51 - 53 years
The Netherlands	1950	70 - 73 years
The Netherlands	1998	75 - 81 years

ad (3) Average healthy life expectancy at birth is only 60 years! This implies that many active workers suffer one or more diseases during their active working life and have an increased risk of disability with advancing age. The concept of the average healthy life expectancy was launched a few decades ago [17]. Average life expectancy (LE) has increased considerably during the past 150 years. However, it is also well-known that much of the gain in years is being spoiled by ill health in old age. Therefore, the need to define a measure of healthy life expectancy (HLE) was felt. A relatively big gap exists between LE and HLE.

HEALTHY LIFE EXPECTANCY AND WORK ABILITY INDEX: WHAT IS THE RELATIONSHIP?

HLE at a certain age is an average figure for the general population. At an individual level this figure may show significant variation between different persons or workers. For actively working people it may be very useful to know their chances and risks of continued ability to work. Therefore, a new concept, that can be called the Working Life Expectancy (WLE), may be useful.

Then the question can be raised whether it might be possible to derive the WLE from the HLE. The link between the two concepts can possibly be obtained from the WAI.

From linear regression analysis on the HLE data in relation to age and the WAI in relation to age, the two following can be derived:

Eq. (1)   Healthy Life Expectancy (HLE) = 57.6 - 0.74 x Age (years)
                    (r = 0.99; expl. variance: 98.7%)
Eq. (2)   Work Ability Index (WAI) = 48.36 - 0.155 x Age
                    (r = 0.29; expl. variance: 8.4%)

The average healthy life expectancy and the average expected work ability [WAI(exp)] can be calculated for each age from equations (1) and (2). For obvious reasons the equation on the WAI is only applicable for the age range between 20 and 65 years.

The following hypothesis can then be proposed: working life expectancy at a certain age [WLE(a)] is the sum of the healthy life expectancy at that age [HLE(a)] and a variable period (dt) depending on the present observed work ability index score [WAI(obs)].

Eq. (3)   WLE(a) = HLE(a) + dt (years)

The period dt is considered the difference between the calendar age of the worker and his functional or physiological age based on the observed WAI score. So,

Eq. (4)   dT = Calendar Age [A(cal)] - Physiological Age [A(phys)]

From eq. (2) and (4) it follows that:

Eq. (5) $A(cal) = [48.36 - WAI(exp)] / 0.155$ (years)
Eq. (6) $A(phys) = [48.36 - WAI(obs)] / 0.155$ (years)
Eq. (7) $dT = [WAI(obs) - WAI(exp)] / 0.155$ (years)

From eqs. (3) and (7) the following equation can be obtained:

Eq. (8) $WLE = HLE + [WAI(obs) - WAI(exp)] / 0.155$ (years)

If WAI(obs) is higher than WAI(exp), then WLE is greater than the average HLE. If WAI(obs) is smaller than the WAI(exp), then WLE is less than the average HLE. The following calculations may illustrate the use of the WLE.

Example:
Male Employee, 50 years; HLE(a) = 20.6 years; WAI(exp) = 40.61. Now, suppose the following two cases:

Case 1: WAI(obs) > WAI(exp); e.g. WAI(obs) = 43
WLE = 20.6 + 2.39/0.155 = 36.0 (years).
So, theoretically this worker is fit to work until the age of 86 years.

Case 2: WAI(obs) < WAI(exp); e.g. WAI(obs) = 38
WLE = 20.6 - 2.61/0.155 = 3.8 (years)

This worker is at risk of early retirement due to diminished work ability at the age of 54 years!

The WLE may be a useful instrument in assessing the risk of future disability of older employees and early retirement from the work force. Further comprehensive analysis of data will be necessary to confirm the value of the concept of WLE.

CONCLUSIONS
Occupational gerontology:
Occupational gerontology may become a useful discipline to study specific aspects of aging and work.

Work ability index:
- The average WAI score showed a significant decrease with age.
- About 14% of the examined employees had a Poor-Moderate Work Ability.

- The Work Ability Index is a useful instrument in secondary prevention of disability. The instrument enables the early detection of downward trends in work ability. These trends can be overcome by appropriate measures for individual or groups of employees.
- The Work Ability Index should be implemented as a standard protocol in Occupational Health.

Working Life Expectancy:

- Working Life Expectancy (WLE) is presented as a new concept.
- The WLE may be useful in assessing the risk of future disability of older employees.

REFERENCES

1. Bergener, M.,1985,: Gerontology - Between Opportunity and Reality. In: *Thresholds in Aging*, edited by Bergener, M., Ermini, M. and Stähelin, H.B., (UK: Academic Press, London), pp. xiii - xix.
2. Ilmarinen, J., 1999, Ageing workers in the European Union, Finnish Institute of Occupational Health, Helsinki, Finland, pp.274.
3. Schulz - Aellen, M-F., 1997, Aging and Human Longevity, Birkhäuser, Boston, USA, 283 + ix pages.
4. Sterns, H.L., Matheson, N.K. and Schwartz, L.S., 1990, Work and Retirement. Gerontology: In *Perspectives and Issues*, edited by Ferraro, K.F. (New York: Springer Publishing Company), pp. 163-178.
5. Ilmarinen, J.,1996, Productivity in late adulthood - physical and mental potentials after the age of 55 years. In: *Aging and Work 3*, edited by Goedhard, W.J.A., (The Hague, The Netherlands: Pasmans), pp. 3 - 17.
6. Kilbom, Å., Westerholm, P. and Hallsten, L.,1997, Work after 45?, National Institute for Working Life, Solna (Sweden), vols 1 & 2, edited by Furåker, B., 299 pages.
7. Ilmarinen, J., 2000, Health problems of ageing workers and their perception of job demands. In: *Aging and Work 4*, edited by Goedhard, W.J.A., (The Hague, The Netherlands: Pasmans), pp. 10 - 18.
8. Ilmarinen, J., 2000, Respect for the Ageing. Concluding Remarks and Reconmmendations. In: *Aging and Work 4*, edited by Goedhard, W.J.A., (The Hague, The Netherlands: Pasmans) pp. 203 - 210.
9. Costa, G., Antonacci, G. Olivato, D. and Ciuffa, V., 2000, Evaluation of functional working capacity by the Work Ability Index in Italian workers. In: *Aging and Work 4*, edited by Goedhard, W.J.A., (The Hague, The Netherlands: Pasmans), pp. 53 - 61.
10. Goedhard, W.J.A., 2000, Experiences with the Work Ability Index in The Netherlands. In: *Aging and Work 4*, edited by Goedhard, W.J.A., (The Hague, The Netherlands: Pasmans), pp. 62 - 67.

11. Monteiro, M.S., Gomes, J.R., Ilmarinen, J. and Korhonen, O., 2000, Aging and work ability among Brazilian workers. In: *Aging and Work 4*, edited by Goedhard, W.J.A., (The Hague, The Netherlands: Pasmans), pp. 68 - 71.
12. Ma Lai-Ji, Jin Xi-peng, Sheng Guang-zu and Wang Xu-ping, 2000, Work ability status of workers in China. In: *Aging and Work 4*, edited by Goedhard, W.J.A., (The Hague, The Netherlands: Pasmans), pp. 72 - 75.
13. Tuomi, K. Ilmarinen, J., Jahkola, A., Katajarinne, L., and Tulkki, A., 1998, Work Ability Index, 2nd edition. FIOH, Helsinki, Finland, pp. 30.
14. Shock, N.W., Greulich, R.C., Andres, R., Arenberg, D., Costa, P.T., Lakatta, E.G., and Tobin, J.D., 1984, Normal human aging: the Baltimore longitudinal study. NIH Publication No. 84-2450, (Washington, U.S: Government Printing Office), 399 pages.
15. Hall, D.A., 1984, The biomedical basis of gerontology. (Bristol, UK: John Wright & Sons), 233 + vi pages.
16. Latten, J. and De Graaf, A., 1997, Fertility and Family Surveys in countries of the ECE Region. Standard Country Report, The Netherland. UN Publications, No. GV.E. 97-0-22, Geneva, Switzerland, 94 + ix pages.
17. Van de Water, H.P.A., Boshuizen, H.C. and Perenboom, R.J.M., 1993, Gezonde en ongezonde levensverwachting (Healthy and unhealthy life expectancy). In *Volksgezondheid Toekomst Verkenning* (Public Health Future Survey), edited by Ruwaard, D. and Kramers, P.G.N., Rijksinstituut voor Volksgezondheid en Milieuhygiene (RIVM), (The Hague, The Netherlands: Sdu Uitgeverij), pp. 203 - 211.

3. Promotion of Work Ability during Aging

Juhani Ilmarinen

Department of Physiology, Finnish Institute of Occupational Health, Finland

ABSTRACT

Aging changes the human resources to work. The contents and demands of work change also remarkable during the life course. The modern concept of work ability describes the interactions between human resources and work characteristics. Work ability declines more often than it improves with age. Therefore the promotion of work ability is needed for the majority of workers. The evidence based promotion concept is based on four dimensions: Age-adjustments are needed in psychosocial work environment and management, in ergonomics and physical work environment, in promting health and functional capacities, as well as in improving the professional competences. Integration of individual and work site actions have been proved to be a powerful combination for employees and workers over 45- years of age. The improvement of work ability has several positive consequences: better productivity, better well-being and better quality of Third Age. Age management, regulation of own work, physical active life style and tailored competence training for older workers are the key actions to be carried out. In the European union, the needs for improvements by Member States have been identified.

Keywords: *Work ability, Promotion concept, Work community, Work environment, Health, Functional capacity, Competence, Age management*

INTRODUCTION

The aging of the work force is a global phenomenon. The main reasons for this historical change are (i) the large baby boom generation born in the1940s-1950s is entering its 50s, and (ii) the subsequent generations have been relatively small. For example, in the European Union (EU) the relative proportion of those aged 50-64 years will be about 27% of the entire work force in 2005. In 2025, this proportion will approach 35%. During the same period, the youngest part of the work force, those aged 15-24 years, will decline, and form only 17% of the EU work force. As a consequence, there will be potentially double the number of older workers as compared to younger workers in the EU over the next 25 years (Eurostat, New Cronos 1998).

The participation rates of aging workers, however, indicate that a dramatic decline occurs after the ages of 50-54 years. In the EU, only 60% of the 55-59 year age group is still working, and the participation rate drops to 20% in the age group of 60-64 years. The major portion of the experienced work force leaves working life several years before mandatory retirement age (Figure 3.1).

One major consequence of their early exit is the worsening of the age dependency ratio (ADR). Today the ADR is about 2.0 in the EU, indicating a ratio of two workers for every non-working person (younger than 15 years, or older than 65 years). The ADR is expected to decline to the level of 1.7 in the year 2025 (Figure 3.2). The lower the ADR, the heavier the economic burden at the societal level. When ADR calculations are done on a more realistic basis (younger than 20 years, older than 60 years), it falls to near 1.0 in several EU Member States in the year 2025. In practice, in this realistic scenario one person is working for every one who is not working. It is understandable that the present quality of life and welfare state services cannot be maintained with such a low ADR.

Therefore, the main solution for keeping the ADR at an appropriate level is to increase the participation rate of the aging work force. For this purpose, integrated and well-coordinated actions are urgently needed at individual, enterprise and society levels. The orientation for aging and work, for responsibilities of different actors and for a variety of actions needed, is illustrated in more detail in Chapter 3 of this article.

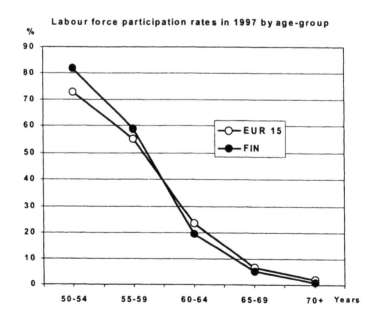

Figure 3.1 Work force participation rates in EU after the age of 50 years

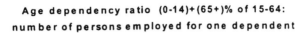

Age dependency ratio (0-14)+(65+)% of 15-64:
number of persons employed for one dependent

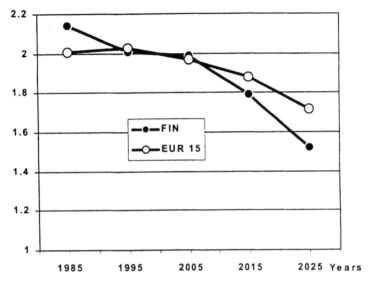

Figure 3.2 Age dependency ratios in EU

CONCEPT OF WORK ABILITY

A basic element for staying at work when older is the individual ability to work. The modern concept of individual Work Ability is illustrated as a house (Figure 3.3). The **Work Ability** house is constructed of four floors. The ground floor is called **Health and Functional Capacities**, where the physical, mental and social dimensions of functionalities are incorporated with health. The second floor is dedicated to **Competence**, including knowledge and skills and the need for life-long learning. The third floor covers the dimensions of **values, attitudes and motivation**. The three floors mentioned describe human resources. The fourth floor describes the **work** as a whole. Both the dimensions of the work community and environment, as well as exposures and work demands, form the content of the fourth floor. Additionally, management is a key characteristic of this floor, indicating that supervisors have the mandate, possibilities and responsibilities in constructing and modifying the dimensions of work.

Therefore, the individual work ability lies both in (i) human resources, and (ii) his/her work characteristics. An appropriate balance between the human resources and work is needed if the individual work ability is to be maintained up to pension age. The model of the work ability house emphasizes that (i) the healthier the ground floor is, the more solid the house, (ii) the second floor interacts with the other two floors, the (iii) the third floor can shake the whole house, and (iv) the fourth floor should not grow larger and heavier independent of the development of available human resources.

All floors change with age, but probably the fourth floor changes more rapidly. Therefore, the new deal is to adjust changes in the work floor to changes in human resource floors. Adjustments are key terms in keeping the house standing. Supervisors are mainly responsible for the work floor and the employees for the other floors.

Figure 3.3 Work ability house

Work ability can be measured by the Work Ability Index (WAI) (Tuomi et al., 1998). Although the WAI was created in the 1980s and the work ability house later in the 1990s, the method works well today and can be described as a culture-free instrument. Reports from Europe, Latin America, Asia and the USA show that the WAI has been feasible in different cultures. The structure of the WAI, as well as its changes with aging, has been reported earlier (Ilmarinen et al., 1997, Ilmarinen 1999).

NEEDS FOR PROMOTION

The main finding in Finnish 11-year follow-up studies indicated that WAI declines with age for 1/3 of the employees, WAI improves for 1/10, and for the majority the WAI remains almost unchanged between 45 and 57 years of age. The decline of WAI is most clear in physically demanding occupations and less pronounced in mentally demanding work. In the range 51-58 years of age, the WAI seems to decline more dramatically than earlier, in all occupations. The prevalence rates of poor WAI (scores 7-27 points) exceeded the level of 25 % of female and male workers aged 58 years in physically demanding occupations. In mentally demanding jobs, the proportion of poor WAI at age 58 varied between 5 and 15%. (Ilmarinen et al., 1997). However, large individual differences existed in all age groups. Aging as such increased the individual differences in all occupations. The main conclusion from the follow-up findings was that promotion of work ability is needed with advancing age, independent of the job. Working alone does not prevent work ability from declining in any studied occupation.

EU work life for aging workers

The EU analysis of working life of aging workers showed that a variety of work characteristics formed a critical burden to workers over 45 years of age. Although large differences exist among the 15 Member States, all countries have neglected the need for several age-related adjustments in both physical and mental work environments.

The country-based analysis summarises the need to improve the situation of the 45+ population in work. The available data was based on the Second European Survey on Working Conditions done in 1995/96. The list of improvements needed in each country has been classified according to individual health and competence issues, as well as exposures in physical and mental work environment characteristics (see Ilmarinen 1999). The countries were ranked according to the existing prevalence rates of 20 risk factors (rank 1=country with a lowest level of risk factor, rank 15= country with highest level of risk factor). Countries ranked worst, e.g., 13-15, by each risk factor were listed and selected as countries requiring the most improvements for their aging work force. A summary of the analysis is shown in Table 3.1.

Table 3.1 Need for improvements by EU Member States

Health and education	EU-country
1. Sickness Absence, high	AUT, DEU, FIN, NDL
2. Work impairment due to chronic diseases, high	AUT, DEU, FIN
3. Physical exercise level, low	LUX, BEL, ITA, ESP
4. Basic education level, low	PRT, GRC, ESP
5. Participation in training, low	LUX, GRC, ESP, IRL
6. Age discrimination, high	AUT, NDL, FRA, GBR, DNK

Physical work environmental	
7. Exposure to vibration, high	ITA, GRC, ESP
8. Exposure to noise, high	FIN, GRC, GBR, ESP, SWE, IRL
9. Exposure to heat, high	AUT, GRC, GBR, ESP
10. Exposure to cold, high	PRT, GRC, ESP
11. Exposure to impurities in air high	FRA, GRC, ESP
12. Exposure to physical load, high	FRA, PRT, GRC, ESP

Mental & social work environment	
13. Use of computers, low	PRT, ITA, GRC, ESP, IRL
14. Tight schedules, often	AUT, FIN, GBR, DNK, SWE
15. Complex tasks, seldom	BEL, PRT, GRC, ESP
16. Learning new things, seldom	PRT, ESP, IRL
17. Regulating own work, seldom	AUT, DEU, GBR, IRL
18. Skills do not match with demands	AUT, LUX, NDL, GBR, SWE
19. Supervisors, lack of co-operation	BEL, PRT, ESP
20. Workhours, heavy schedules	PRT, ITA, ESP, IRL

Aging and Work: Orientation Matrix

The theme of aging and work is a very complex one. Therefore, a matrix has been created to cover different aspects of aging and work, which should be taken into consideration when the situation of aging workers is to be improved.

The matrix has been described elsewhere (see Ilmarinen 1999, Ilmarinen 2001a). It combines problems, solutions and aims at three levels: individual, enterprise and society. The items indicated in the nine boxes of the matrix have been taken from research projects or policy reports. The main message of the matrix is that we must work at all three levels when aiming to improve the situation of aging workers. Problems should be identified first, aims can be determined and solutions for achieving the aims should be selected.

According to Finnish experiences, the enterprise level is most important because the needs of both individual and enterprise can be met together and at the same time. Society level regulations are important too, but their development often requires several years before implementation. Enterprises, however, have a sufficient degree of freedom to carry out the improvements needed without new legislation. Because most societies do not have too much time to solve the problems of baby-boomers reaching their 50th birthday, the most effective and fastest responses for aging can be realised at the company level.

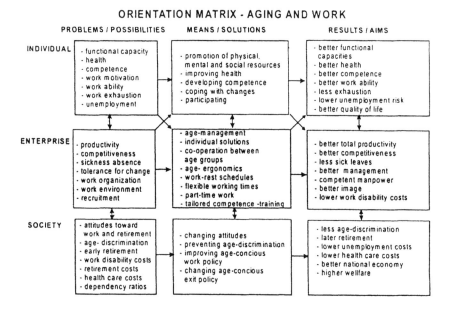

Figure 3.4 Aging and work: Orientation Matrix

PROMOTION CONCEPT FOR WORK ABILITY

Based on the follow-up research in 1980s and on the FinnAge Respect for the Aging Programme (1990-1996) a promotion concept of work ability was constructed in the Finnish Institute of Occupational Health (Tuomi et al., 1997, Ilmarinen and Louhevaara 1999, Ilmarinen and Rantanen 1999). The basis for the concept was the scientific evidence in identification of factors improving and reducing work ability during aging (Tuomi et al., 1997). The concept has been tested in several research and developmental projects. The validity of the model has been tested recently (Tuomi et al., 2001).

The concept has four dimensions, all of which are needed for the promotion of work ability (Figure 3.5). The concept also includes the side effects of the actions. The improvement of work ability is followed by improvement of work quality and productivity, individual life quality and well-being, as well as long-term effects on Third Age quality. Short descriptions of the different dimensions are given here. For further reading, several articles are available (Ilmarinen and Rantanen 1999, Ilmarinen 2001a, Tuomi et al., 2001).

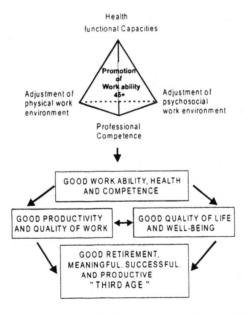

Figure 3.5 Concept for the promotion of work ability

Dimension 1: Work community and management

Adjustments for psychosocial work environment include e.g., age-management, flexible working time schedules and teamwork concepts. Age-management is probably the most powerful tool for improvement of work ability of older workers. Therefore age-management will be introduced in more detail in Chapter 5.

Dimension 2: Physical work environment

Adjustments needed in work content and in physical work environment refer to ergonomics. Age-ergonomics include e.g., the decline of physical workload, work-rest schedules, and possibilities to regulate own work. Because the possibilities to regulate own work is essential for aging workers, this topic will be described in more depth in Chapter 5.

Dimension 3: Health and functional capacities

Traditional health and lifestyle promotion is composed of good practice by occupational health services as well as personal lifestyles supporting good health and well being. From a large variety of lifestyles, regular physical exercise exceeds in importance some other risky lifestyles (nutrition, smoking, alcohol consumption, etc.) among aging workers. However, an important topic is the change in health status and health adjustments needed at work. This topic will be described more in Chapter 5.

Dimension 4: Professional competence

Updating skills and knowledge is a necessity in the majority of jobs. New work ability demands concern both young and older workers. In particular, learning new skills in IT technology often creates burdens for older workers. It should be pointed out, however, that learning is not dependent on working age, but on the way the learning is organised. Older people learn differently than younger ones. Therefore, new training methods are needed for aging workers.

Integration of four dimensions: a powerful combination

The lesson learned in the 1990s was that to make the concept powerful, the four dimensions should be integrated. In real life, however, some priorities should be set first, because starting four processes at the same time is seldom feasible at company level. To prioritize the dimensions needed, an analysis of their importance should be made. In principle, starting with the easiest and most acceptable one is to be recommended. It should be emphasized, however, that in most companies processes in all four dimensions are needed; so the aim is to get these processes started up over a period of 1-2 years. Experiences show that an appropriate time for adjustments needed at the work site and in individual resources is about three years. Depending on the culture of the company, the actions for all dimensions should be carefully tailored. A good promotion plan is needed, including evaluation of processes and results. In the long run, the processes for promotion should be integrated in the normal daily routines of each worker and foremen. Promotion of work ability should be a part of individuals' ordinary daily activities at work, not a special project with a strict timetable.

Productivity, well-being and Third Age

The improvement of work ability has several positive consequences. In general, the cost-benefit analysis over 200 promotion projects in Finland showed that the ratio varied between 3 and 20; the investments came back to the company at least three-fold, and in the best cases even 20-fold. The benefits were most pronounced in the decrease of work disability costs and of sickness absence, and in the increase of productivity.

The follow-up study showed that good work ability was statistically significantly related to high quality and high productivity in one's own work. When the group with excellent work ability was compared with the group with poor work ability, high quality of work and high productivity were >1.5-fold more common among the former, the corresponding figures for high quality of work alone and high productivity alone being >1.9- and > 1.3-fold, respectively (Table 3.2) (Tuomi et al., 2001).

Table 3.2 Associations of work ability with the quality and productivity of one's own work (%) among people 55-62 years of age in 1992

Class of work ability	n	HQ/HP[a]	HQ	HP	Other accounts	p[b]
Excellent	105	15,2	28,6	21,9	34,3	<0.001
Good	269	21,5	28,2	9,7	40,5	
Moderate	466	10,7	20,8	15,5	53,0	
Poor	176	9,6	14,8	16,5	59,1	

[a] HQ, high quality of work; HP, high productivity
[b] P value, based on x^2 test

The relationship between work ability and well being and quality of life were examined, too. The work ability index explained the opinion of retirement as being best ($R2=0.30$, n=1005), with enjoyment of staying in one's job ($R2=0.23$, n=1015), and life satisfaction ($R2=0.13$, n=1012) coming next. According to these results, the quality of life and well-being were significantly better among those whose work ability index was good than among those whose work ability index was poor (Tuomi et al., 2001).

The predictive power of the work ability index five years earlier was examined among the retired subjects in 1997. The predictive power of the work ability index was highest for subjective estimation of current work ability compared with lifetime best, physical condition, perceived health, and life satisfaction. The ability to function and well being upon retirement were significantly better in the groups whose work ability index had been good or excellent five years earlier than among those whose work ability index had been poor or moderate (Table 3.3) (Tuomi et al., 2001).

Table 3.3 Proportions (%) of good functional ability at retirement in 1997
according to the work ability index, as measured in 1992

WAI	Good retirement ability in 1997			
in 1992	good WAI n=619	good health n=700	good PWC n=700	satisfied with life n=704
excellent	63,3	73,3	80,0	35,0
good	56,9	62,8	65,6	29,9
moderate	22,5	36,6	38,8	20,5
bad	5,7	20,0	19,6	8,5
x^2	178,3	138,5	145,3	49,5
p	0,001	0,001	0,001	0,001
df	30	12	12	12

EXAMPLES OF ACTIONS NEEDED FOR AGING WORKERS

In the following, one example of actions needed is introduced in each of the four dimensions of the promotion concept.

Work community: Age Management

Age management training focuses on skills and knowledge for all levels of managers like foremen, supervisors, and line managers as well as top managers. The basic question is: what should managers know about aging? The training concept is based on the (i) lack of information about aging in managerial training programmes, and (2) the correction of several myths about work force aging. The aim of Age Management training is to change attitudes toward aging by dissemination of facts.

The list of topics in Age Management training is given in Table 3.4. The general information of society level and company statistics will be followed by work force participation rates and dependency ratios. The modern concept of work ability is an essential part of knowledge, making the promotion of work ability possible. It is followed by aging and economics, where e.g., the myth of aging and productivity will be corrected: productivity is not dependent on age but on the organisation of work.

Relevant information about aging and health brings up the issue of adjustments needed at work when health status changes. The work must be organised so that the two-thirds of employees aged 55+ having diagnosed diseases can continue working. The knowledge of mental growth is the strongest evidence of positive changes with age. Here the challenge is to identify items at work that should grow in the same direction the aging workers are growing. Additional information is needed to describe the relationship between physical and mental capacities and social functioning.

Aging and learning is one of the key issues in modern working life because of the growing and new work demands which are common for aging workers, too. The myth that "an old dog cannot learn new tricks" must be corrected. Learning as such is not dependent on age, but older people learn differently than do younger ones. The five differing learning characteristics between older and younger generations emphasize that e.g., training for older workers must be tailored and organised correctly.

Finally, Age Management training will focus on prevention of overload at work, introducing the age management toolboxes, and prevention of age discrimination. Special attention will be given to the basic toolbox. Four tools cover the (i) calibration of the age attitude, (ii) co-operation and support of teams combining young and old, (iii) identification of individual needs at work, and (iv) improving communication skills. All four of these items have been shown to increase the work ability of aging workers. The basic tools have been introduced in more detail elsewhere (see Ilmarinen 1999).

Table 3.4 Topics for Age Management training

- Age structure of the company today and in the future
- Participation rates, dependency ratios
- Modern concepts of work ability
- Work ability and employability
- Work ability and economics
- Aging and productivity
- Aging and health
- Aging and mental growth
- Aging and functional capacities
- Aging and learning
- Regulation of work load
- Age-management Tool-box
- Prevention of age discrimination

Work environment and ergonomics: Regulating own work

Aging workers need possibilities to regulate their own work. The basic reasons for this need are: (i) changes in functional capacities, (ii) increased individual differences in capacities, and (iii) increased recovery time from workload. Breaks at work, order of work tasks, working methods and working speed are the key issues for regulation.

The possibilities to regulate these four characteristics of work have been summed up using ranking analysis by Member States of the European Union. Large differences can be seen between the countries (Table 3.5). The best possibilities for regulating work were identified in Denmark and Portugal, and the worst possibilities in Austria and Germany. In general, the possibilities were better for older men than for older women, the exceptions being Germany, Austria, Greece, France and especially Spain. A notable imbalance in the other direction (control of women worse than that of the men) occurred in Denmark, Sweden, England and Ireland. The Nordic countries are normally considered to have the most equally balanced societies, but this is not the case in regulating their work. Finland was the most balanced of Nordic countries but its ranking order was too far from the top. Generally, the results show that many

European countries have a rigid working life for older workers (Ilmarinen 1999).

Probably the most important need for older workers is the possibility to regulate breaks at work. About 70% of men and 60% of women over 45 years old were able to take break when they wished, and in most EU countries these possibilities were better for older than for younger men and women. This finding is a positive one, because the recovery from workload increases with advancing age. An optimal work-rest schedule can prevent the cumulative increase of fatigue, physical and mental, during the working day. However, more research is needed to give guidelines and recommendations according to age and workload.

Age-ergonomics covers a variety of practical issues that should be considered both in work sites and in individual behaviour. A list of 13 topics has been introduced by Spirduso (1995).

Table 3.5 Possibilities to regulate own work

Member state[1]	men	women	all
DNK	8	22	30
PRT	15	15	30
SWE	17	31	48
LUX	23	25	48
BEL	24	26	50
NDL	24	26	50
ITA	26	34	60
ESP	53	12	65
FIN	32	35	67
FRA	42	31	73
GRC	43	33	76
GBR	31	47	78
IRL	31	49	80
AUT	53	40	93
DEU	58	54	112

[1] ascending order according to all

Health and functional capacities: Health adjustments at work due to chronic health problems

Because the number of chronic diseases increases with age, some adjustments at work are often needed, too. In the EU about 20-25 % of aging men and women felt that their job became more difficult because of a chronic or permanent health problem. Men and women over 45 years of age felt so more often than younger ones. The differences between the countries were remarkable: the job became more difficult due to chronic health problems about three times more often among aging men in Austria, Germany and Finland than in Spain, Denmark and Sweden; aging women in

Austria, Greece and Germany felt this about three times more often than women of same age in Spain, England and Sweden (Ilmarinen 1999).

Country differences are difficult to explain and the results are always culture-dependent. In Finland, chronic diseases hinder the work of men more than that of women, and the situation worsens more markedly among men than among women with age. The degree to which the more severe cardiovascular diseases of the male population after the age of 45 years affect these results remains to be studied in further research.

The difficulties that are caused by chronic diseases could be eased, especially by arrangements at the workplace, and work tasks and methods could be changed to better correspond to the state of health of aging workers. Better treatment for diseases is necessary, but it does not make adjustments at work unnecessary. Because the adjustments needed are individual and dependent on the chronic disease, the role of occupational health services as an expert organisation is essential. The adjustments can be identified and realised by a group consisting of the worker, OHS and the foreman.

General guidelines and recommendations for adjustments should be worked out for most common chronic diseases of aging workers. However, it is always an individual matter and must be tailored according to the rights and possibilities of employee and employer. It should be pointed out that the one-third of workers over 45 years requiring adjustments constitutes a major portion of the work force. Their possibilities to continue working should be guaranteed by appropriate adjustments.

Health promotion through healthy lifestyles is a powerful preventive tool for aging workers. Recent findings emphasize the role of regular physical exercises (Blair et al., 1996). In general, an active lifestyle in the physical, mental and social senses is a good concept to prevent premature decline in functional capacities (Ilmarinen 2001b).

Professional competence: learning new things

Aging workers learn differently than younger workers. The organization of learning should take into consideration the following aspects:

- learning strategy
- learning conditions
- learning methods
- learning speed

The learning strategy emphasizes the role of experience. A new issue should be connected with existing knowledge. In this way, issues are learned in more depth and the conception of new things improves. Older employees are self-guided (i.e., they know what they want to study and how). The teacher acts primarily as a counselor.

Learning conditions should be free of additional disturbances, because older employees are more sensitive to extra stimuli. Extra noises through the auditory sense overload the filters, which should distinguish between the important and non-relevant information. An auditory mismatch disturbs the learning processes.

Mechanical external memory is weakened, but the deficit can be compensated for by connective memory. New and unfamiliar things are connected to familiar ones. New information is brought into relation with the existing information structure. It is important to fix the new information through the work. With the help of imagination, the meaning of new information can be understood to be positive for work performance: it makes work performance easier and better. It is worth learning.

A natural means of learning new things for aging workers is learning by doing. It emphasizes an active, practical view of learning. Therefore, the time for learning is longer for aging workers. In learning IT technology, the practical aspect often takes more time than the theoretical aspect. Understanding new terms and finding information from user's manuals is slower for aging workers.

Based on the differences in leaning between the generations, the courses should be tailored for aging workers. Also, the role of teacher is changing: the teacher should act as a supporter of the learning, not as a traditional teacher. Therefore, teachers of aging workers should update their teaching skills.

More information about aging and learning in working life is available (Ilmarinen 1999).

AGING AND SUSTAINABLE WORK ABILITY: KEY ELEMENTS

This article has described evidence-based information in aiming to promote work ability during aging. A sustainable work ability is the goal for everybody in the working life. It can be achieved by appropriate means. The key elements in this article can be summarised as follows:

1. Work community
    *   age management
    *   individual needs
2. Work environment
    *   regulation of own work
    *   work-rest schedules
3. Health and functional capacities
    *   ill-health adjustments
    *   active lifestyle
4. Professional competence
    *   new training concepts
    *   teachers training
5. Integration of measures

Promotion of work ability is the basis for employment with advancing age. Promotion of work ability, however, should be combined with employability activities. Together these comprehensive processes can change individual attitudes, the atmosphere in enterprises and the rules of societies towards a better world for aging workers. However, the fight for the aging is also a fight for the young - they, too, should be ensured a good future.

REFERENCES

1. Blair S.N., Kampert J.B., Kohl H *et al.*, 1996, Influences of cardiovascular fitness and other precursors of cardiovascular disease and all-cause mortality in men and women. *JAMA 276*, pp. 205-210.
2. Eurostat. New Chronos. Theme 3, 1998, Population and Social Conditions. (CD-ROM).
3. Ilmarinen J., Tuomi K., and Klockars M., 1997, Changes in the work ability of active employees over an 11-year period. *Scand J Work Environ Health 23: suppl 1*, pp. 49-57.
4. Ilmarinen J. (ed.), 1999, Aging workers in the European Union- Status and promotion of work ability, employability and employment. Finnish Institute of Occupational Health, Ministry of Social Affairs and Health, Ministry of Labour. Helsinki.
5. Ilmarinen J., Louhevaara V. (eds.), 1990-1996, FinnAge - Respect for the aging: Action programme to promote the health, work ability and well-being of aging workers.1999. *People and Work, Research reports 26*, Finnish Institute of Occupational Health, Helsinki.
6. Ilmarinen J. and Rantanen J., 1999, Promotion of work ability during aging. *American Journal of Industrial Medicine Supplement 1*, pp. 21-23.
7. Ilmarinen J., 2001, Functional capacities and work ability as predictors of good 3rd age. In: *Physical Fitness and Health Promotion in Active Aging*, edited by K. Shiraki, S. Sagawa and M. Yousef, (Leiden: Backhuys Publishers), pp. 61-80.
8. Ilmarinen J., 2001, Aging workers. *Occupational & Environmental Medicine*, Vol 58, No 8, pp. 546-552.
9. Spirduso W.W., 1995, Job performance of the older worker, In: *Physical dimensions of aging*, Chapter 13. edited by W.W. Spirduso, (Champaign, Ill: Human Kinetics), pp. 367-387.
10. Tuomi K (ed.), 1997, Eleven-year follow-up of aging workers, *Scand J Work Environ Health 23*, **suppl 1**.
11. Tuomi K, Ilmarinen J and Jahkola A *et al.*, 1998, Work Ability Index. *Occupational Health Care* **19**. Fnnish Institute of Occupational Health, Helsinki.
12. Tuomi K, Huuhtanen P, Nykyri E and Ilmarinen J., 2001, Promotion of work ability, the quality of work and retirement. *Occup. Med. Vol. 51, No. 5*, pp. 318-324.

4. Terminology of Aging Used in Legislation and Governmental Policy

Seichi Horie, Takao Tsutsui

*Department of Health Policy and Management,
University of Occupational and Environmental Health, Kitakyushu, Japan*

ABSTRACT

This study describes the international similarities and differences of both the policies formulated for the aging society and the vocabulary associated with age. Public literature of the UN, OECD, and eight countries were examined. Increased life expectancy and low fertility in the studied countries were regarded as major causes of the aged society at present. Baby-boomers, economic recession and cultural background affected the aging of the studied countries differently. Policies relating to the pension crisis, unemployment and age discrimination were interrelated. Mandatory retirement was legally prohibited in the USA, Canada and New Zealand. The importance of occupational safety and health in the aged society can never be overemphasized, even if the issue is not exactly stipulated in legislation. International standardization and cooperation are concerns for the future. The most widely prevalent definition of aged people implies people of age 65 or over. Related terms such as "aged," "old," "elderly," "young-old," "old-old," and "middle-aged" are used with different terminology across the countries. Clear definitions were applied in policies on pension and insurance, although not necessarily in labor policies. The notion of "aged worker" was generally younger than that of "aged people." Terminology should be mutually understood and clarified at the level of international communication.

Keywords: *Terminology, Aging, Aged worker, Middle-aged, Elderly worker, Older worker, Policy, Legislation*

INTRODUCTION

Since the late 20th century, many countries have struggled to cope with the aging of the work force [1]. Though age-associated biological changes occur in relation with age, there exists wide individual difference [2,3]. Therefore, today, we generally agree that setting limit at certain age or making judgment by age has little sense for the scientific evaluation of individual ability to work [4]. Moreover, people blame those attitudes to limit one's opportunity as the discrimination by unproved prejudice with age.

Yet, in the process of making policies, rules and laws, we inevitably mention the vocabulary associated with age. Governments and organizations apply a variety of terms such as "aged people," "aged worker," "middle-aged," "old" and "elderly," with diverse definitions. These words often appear in official statements for social security schemes, stabilization of employment, and health promotion [5]. However, we are often confused by the lack of clarity about the meaning of these words.

This study was conducted to describe the international similarities and differences in both the legislation and policies formulated for the aging society and the definition of officially adopted vocabulary associated with age.

MATERIAL AND METHODS

We searched articles and publications that bear keywords of "ag(e)ing," "aged worker," "elderly worker" and "older worker," listed in OSH-ROM computer database (SilverPlatter Information Inc.) and materials in the National Diet Library of Japan. We also collected public statements of United Nations (UN), Organization for Economic Cooperation and Development (OECD), United States of America (USA), the United Kingdom, Canada and New Zealand, as well as English translations of legislation of France, Germany, Sweden and Japan, regarding employment, social security and health. We made lists of policies and terminology used in those statements with their age definitions. Then we reorganized the collected information to compare them and to summarize the interrelation of policies on aging.

RESULTS

Population and Employment

The current rapid aging of society has never before been seen in global history. Increasing life expectancy and decreased fertility are the two primary reasons for the aging of society. Aging of postwar baby-boomers also causes a drastic change in population pyramids. The number of people aged 60 or more and their proportion to the whole population increased from 200 million and 8% in 1950 to 600 million and 10% in 2000. It will reach 1 billion and 13% in 2020 and the most serious scenario predicts 2 billion and 20% in 2050. The UN has proclaimed that each country should prepare for aging since the 1980's. The labor market suffers from a surplus of aged workers and depletion of young workers. Aged workers have less opportunity to work, they may not choose appropriate work for their physiological and cognitive capacities, and during economic recession they are also pressed to make way for younger workers.

It took only 24 years (1970-1994) for the age composition of Japan to increase the proportion of people aged 65 or over from 7% to 14%; the same change took 115 years (1865-1980) in France. In 2000, 22.3 million citizens were age 65 or over, composing 17.5% of the total population. This figure will reach 20% in 2006 and 27% in 2020. The proportion of those aged 60 or over among all workers will reach

20% in 2020. Systematic reforms of the pension system, health insurance, and employment remain as the urgent issues to be solved.

Human Rights and Employment

Human rights should be highly respected at work. The question is whether a decision based on age is discrimination. The ILO Convention concerning Discrimination in Respect of Employment and Occupation of 1958 (C111) requires equality of opportunity and treatment, and elimination of any discrimination, made on the basis of race, color sex, religion, political opinion, national extraction or social origin, in respect of employment and occupation. Article 5 of C111 refers to age in the phrase saying that protective measures generally recognized from reasons such as sex, age, disablement, family responsibilities or social or cultural status shall not be deemed to be discrimination. The Termination of Employment Convention of 1982 (C158) listed 10 invalid reasons for termination of employment at the initiative of the employer: race, color, sex, marital status, family responsibilities, pregnancy, religion, political opinion, national extraction or social origin. However, age was not included there. The Older Workers Recommendation of 1980 (R162) and Termination of Employment Recommendation of 1982 (R166) proclaimed that age should be considered as an invalid reason for discriminating in employment, selection of occupation and retirement.

In the course of global movement toward the International Year of Older Persons in 1999, the UN addressed five principles to be followed: independence, participation, care, self-fulfillment and dignity of older persons. The UN facilitated the elimination of age discrimination at work, advocating that older persons should remain integrated in society and should develop opportunities for service to the community in positions appropriate to their interests and capabilities. According to the fourth worldwide survey on the International Plan of Action on Ageing performed by the UN in 1996, 33 of the 55 responding countries (60%) had a fixed retirement age. The proportion was gradually decreasing in the world, especially among developed nations. The actual age of retirement widely varies across the countries (Table 4.1) [6]. It also depends on gender and sector.

Table 4.1 Average age of retirement* in OECD countries

country	male			female		
	1960	1980	1995	1960	1980	1995
Iceland	68.8	69.3	69.5		65.8	66.0
Japan	67.2	67.2	66.5	64.6	63.9	63.7
Switzerland	67.3	65.5	64.6	66.9	62.4	60.6
Norway	67.0	66.0	63.8	70.8	61.5	62.0
Portugal	67.5	64.7	63.6	68.1	62.9	60.8
Turkey	68.7	64.9	63.6	69.2	67.6	66.6
USA	66.5	64.2	63.6	65.1	62.8	61.6
Ireland	68.1	66.2	63.4	70.8	66.0	60.1
Sweden	66.0	64.6	63.3	63.4	62.0	62.1
Denmark	66.7	64.5	62.7	64.6	61.0	59.4
United Kingdom	66.2	64.6	62.7	62.7	62.0	59.7
Canada	66.2	63.8	62.3	64.3	60.5	58.8
Greece	66.5	64.9	62.3	64.4	62.5	60.3
New Zealand	65.1	62.9	62.0	62.5	58.7	58.6
Australia	66.1	62.7	61.8	62.4	58.2	57.2
Spain	67.9	63.4	61.4	68.0	63.6	58.9
Italy	64.5	61.6	60.6	62.0	59.5	57.2
Germany	65.2	62.2	60.5	62.3	60.7	58.4
France	64.5	61.3	59.2	65.8	60.9	58.3
Finland	65.1	60.1	59.0	63.2	59.6	58.9
Netherlands	66.1	61.4	58.8	63.7	58.4	55.3
Austria	63.9	60.1	58.6	61.9	59.3	56.5
Luxembourg	63.7	59.0	58.4	63.8	60.8	55.4
Belgium	63.3	61.1	57.6	60.8	57.5	54.1

(years old)

* Calculated from participation rate of quinquennial age group under the
assumption that retirement starts at age 45 and increases linearly.

Source: OECD study on the policy implications of ageing, 1998

In Japan, the fixed-age retirement system combined with the long-term employment contract has spread in society since the early 20th century. On the contrary, in Western European and North American countries, it is regarded as unlawful to discriminate against a person merely because of his/her age with regard to multiple conditions of employment, including recruitment, hiring, training, assignment, transfer, promotion, wage, compensation, disability leave, benefits, retirement plans and firing.

The USA takes a typical position, by clearly abolishing age discrimination. The Age Discrimination in Employment Act of 1967 protects individuals who are 40 years of age or older from employment discrimination based on age. It applies to private employers with 20 or more employees and state and local governments. Therefore, even at recruitment, it is unlawful either to include age preferences, limitations, or specifications in job notices or advertisements, except in the rare circumstances where age is shown to be a "bona fide occupational qualification" reasonably necessary to the essence of the business. The Older Workers Benefit Protection Act of 1990 specifically prohibited employers from denying benefits to older employees. Today, in the same context, the mandatory retirement age of 60 set for airplane pilots by the Federal Aviation Administration is currently under discussion. Bills presented in the Senate required they should be judged on their individual ability and skills and criticized the present rule in that it is based on unproved assumptions about age. Thus, in the USA, the terminology of age itself can barely be mentioned in the labor contract. Moreover, disability also does not qualify as a criterion to select workers. The Americans with Disabilities Act of 1990 (ADA) prohibits private employers with 15 or more employees and governments from discriminating against qualified individuals with disabilities in job application procedures, hiring, firing, advancement, compensation, job training, and other terms, conditions and privileges of employment. The qualified individual with a disability is defined as a person who has a health impairment that substantially limits major life activities, but who can perform the essential functions of the job in question with or without reasonable accommodation. Employers are required to provide reasonable accommodations such as making existing facilities used by employees readily accessible to and usable by persons with disabilities, job restructuring, modifying work schedules and reassignment to a vacant position, unless those actions require significant difficulty or expense to the employer.

In New Zealand, the Human Right Law of 1993 prohibited discrimination against people on the basis of ethnicity, religion, disability, political belief, age and gender, on the subject of employment. This means that mandatory retirement was abolished there.

In France, it is prohibited to advertise an age limit or preference at job recruitment and to terminate the labor contract because of age. However, it is interesting that a major trade union in Paris has an employment treaty stating that, if the worker is at the pensionable age of 65 or over, the dismissal of the worker should be regarded as voluntary resignation.

In Sweden, Section 23 of the Employment Protection Act describes a similar policy to the ADA in the USA, stating "An employee who has reduced working capacity and who has, therefore, been given special duties by the employer shall be given priority for continued work ... where such can be accomplished without serious inconvenience to the employer." Section 33 of the Employment Protection Act stipulates "An employer desiring an employee to leave his or her employment when the employee reaches the age at which retirement with old-age pension becomes compulsory ... shall give the employee at least one month's written notice of such." This implies legally admitted termination of employment at the employer's initiative with regard to the pensionable age limit.

Pension and Employment

Until the 19th century, when the pension scheme was not well established in the world, retired people without family support were either rich enough from property income or had to live in poverty. After the public pension expanded, people became able to retire early and also to live longer. When workers decide the timing of retirement, they are concerned with both the pensionable age and the term of unemployment insurance benefit. Adoption of the pension system as the intergenerational support mechanism in society caused a decline in work participation rates after pensionable age, to the extent of 20% in the USA and 10% in Europe in the 1970's. In the 1980's, some European countries introduced a policy called work sharing, specifically designed to encourage older workers to decide to withdraw from work earlier, and at the same time to receive pension benefits earlier, so as to make room for the younger generation to work. The policy against unemployment took precedence over that for the pension system in those days. Yet in the 1990's, the aging society forced established pension plans to increase their contributions and to reduce benefits. The pensionable age of the earning-related public pension plan of OECD countries is summarized in Table 4.2 [6]. In some countries, people may choose early or delayed payment plans under special conditions. Many programs are currently under reform.

Table 4.2 Pensinable age for public earning-related program** in OECD countries

country	male early←standard→deferred			female early←standard→deferred		
Iceland	65	67		65	67	
Norway	64	67	70	64	67	70
Denmark	60	67		60	67	
Ireland	55	66		55	66	
USA	62	65 (67)		62	65 (67)	
Canada	60	65	70	60	65	70
Finland	60	65	no limit	60	65	no limit
Sweden	61	65	no limit	61	65	no limit
Luxembourg	57	65	68	57	65	68
France	60	65		60	65	
Spain	60	65		60	65	
Mexico	60	65		60	65	
Netherlands		65			65	
Portugal	60	65		60	64	
Switzerland	63	65	70	60 (62)	62 (64)	67
Australia	60	65	70	60	61 (65)	66
Belgium	60	65		60	61 (65)	
Germany	60 (62)	65	no limit	(62)	60 (65)	no limit
United Kingdom	50	65	70 (no limit)	50	60 (65)	70 (no limit)
Poland	60 (62)	65		55 (62)	60 (65)	
Greece	60	65		55	60 (65)	
Austria	60	65		55	60	
Italy	52	63	65	52	58	65
New Zealand		62 (65)			62 (65)	
Korea	55	60 (65)	65	55	60 (65)	65
Japan		60 (65)	70		59 (65)	70
Czech	57	60 (62)		54	57 (61)	
Slovak	58	60	no limit	55	57	no limit
Hungary	(60)	60 (62)	no limit	(55)	56 (62)	no limit
Turkey	43	55		38	50	

** Number based on data in 1997. Some programs are combined with unemployment insurance system. Multiple exceptions and variations exist. Numbers in parenthesis are those for future reform.

Source: the OECD study on the policy implications of ageing, 1998

In France, standard pension benefit starts from age 65; however, certain retired workers may claim it from age 60 at a reduced rate. Meanwhile, a policy called ARPE (allocation de remplacement pour l'emploi) facilitates replacing the labor force with the younger generation from the age 50, by collecting funds called Delalande Contribution for ASSEDIC (Associations pour l'emploi dans l'industrie et le commerce).

In Germany, standard pension benefits start from age 65 for males. In 1972, a special pension benefit from age 60 was introduced for workers after unemployment. However, unexpectedly large number of workers retired from companies after the age of 57 years old and 4 months to claim the unemployment benefit, which continues for 32 months at the most, then moves to a special pension benefit at age 60. The Pension Reform Act of 1992 raised all of these pension age thresholds to 65. Meanwhile, the Act Promoting Gradual Transition into Retirement of 1996 enabled workers at the age 55 or over to take partial retirement to create partial labor vacancy. For example, two older workers may perform one worker's task at a 70% salary, including a 20% supplement from unemployment insurance. In this way, they seek the coordination of intergenerational sharing of wealth and employment.

In New Zealand, the Law on Pension for Aged People was enacted as early as 1898, targeting age 65 or over. In 1977, the National Superannuation Law was proclaimed for people age 60 or over. In 2001, the age limit was returned to 65 years old.

In Sweden, a public pension system was introduced in 1914, covering people at age 67 or over and disabled people. In 1960, the pay-as-you-go (PAYG) system was introduced to create an income-related scheme. Then, the standard of pensionable age was reduced to 65. Those who decided on early retirement became entitled to receive benefits from 61 years old with a reduction of 0.5% for every month prior to age 65. Meanwhile, the Employment Protection Act of 1997 ensured people the right to work until age 67, and people unemployed for more than 6 months at age 55 or over may enter a public employment program for older people. The unemployment insurance benefit may be claimed at 80% of original income for 450 days maximum, starting from age 57. The National Action Plan for the Elderly of 1998 introduced a new National Old-age Pension and removed the upper age limit from the pension bonus, which gives 0.7% maximum for each month after age 65 to enable people who wish to work to do so, with the intention of raising the work participation rate of aged people.

In Canada, the Old Age Security Act of 1927 covered those age 70 or over, not strictly limited to Canadian citizens. In 1965, the standard pensionable age was lowered to 65. People may also adjust their retirement age between 60 to 70. If people start the pension early, it is permanently reduced by 0.5 percent for the number of months before age 65. If people start later, it is increased by 0.5 percent, in the same way. The Guaranteed Income System provides a monthly benefit to low-income people between the ages of 60 and 64.

In the USA the original Social Security Act was enacted in 1935. The standard age for starting to receive pension benefit from OASDI (Old-Age, Survivors and Disability Insurance) will be raised from 65 to 67 by 2027. Qualified workers may claim for old age insurance from age 62 at a reduced rate. In contrast, people can save it until the age of 70 to receive an additional amount.

Health and Employment

Health insurance systems for aged people are also under reform in many countries. In the USA, Medicare currently starting at age 65 will be raised to age 67 by 2007. Similarly in Japan, the Health Insurance Law and Medical Service Law for the Elderly were amended in 1997 to raise the insurance premium and co-payment, and raising the starting age from 70 to 75 is now under discussion.

Aged people who remain in the labor market often have to change their workplace. If their working conditions are left unmodified, their physiological characteristics and the heterogeneity of the work force may become a safety and health issue.

For the International Year of Older Persons in 1999, the UN proclaimed that, when employing older persons, characteristics of working conditions, environment, scheduling and organization of work should be taken into account. To prevent occupational diseases, it is recommended to perform re-education, pre-retirement medical checks, and gradual transition from active working life to retirement with lightened workload during the last years of the working life, for example by gradual reduction of work-time.

In Japan, Article 62 of the Industrial Safety and Health Law stipulates that in respect of middle-aged and aged workers and others to whom considerations should specially be given in placing them with a view to preventing industrial accidents, the employer shall endeavor to see that they are assigned to an appropriate job according to their physical and mental conditions.

In New Zealand, renewal of driving license requirements for people at age 75 and over requires regular health and vision examination and evaluation of driving skills, for the safety of both the driver and other road users. In Sweden, the Work of the Elderly Policy Project was continued until 2000 to cope with the issue. In Japan, the new national health promotion movement called Healthy Japan 21 is currently under way, with numerical targets until 2010 to decrease premature death, defined as death before age 65, and to extend the period during which people can live without suffering dementia or being bed-ridden.

Terminology of Aging

Official statements, surveys, legislation and policies have defined age-related terms in multiple ways (Table 4.3). We observed some general patterns in those definitions. First, age directly implies duration of years the person lives after birth and rarely involves the notion of functional ability or actual activity in society. Second, when the exact age periods had to be defined, the duration started at the number of years multiplied by 5 years such as 60, 65 or 70. Third, adjectives such as aged, old, elder and senior are often used in the texts discussing population, pension, insurance, revenue and discrimination, but not so frequently in those for employment and occupational health. Fourth, the term "old-aged people" generally meant retired persons, whereas "old-aged workers" were literally working, and there are decades of actual age difference between these two terms. Fifth, definition of age slides towards an older age according to the aging of the target population. Incidentally, the oldest category adopted in the summary table of the census in the USA was 85 or over, while Japan has a category of 100 or over. Sixth, analytical documents sometimes use additional vocabulary to express the diversity among old-aged people: young-old for people aged 65-74 years old, middle-old for 75-84, old-old or oldest-old for 85 and over. UN statistics for world population adopted the expression of "very old people" to categorize those aged 80 and over. Finally, many documents avoid defining when people enter or leave the period of young-age, middle age, or old age.

Table 4.3 Terminology and definition of age used in policy, legislation and official survey

policy, legislation and survey	age or age range
health policy in Japan	
old people	65+
older old population	75+
younger old population	65-74
old age	65+
middle age	45-64
manhood age	25-44
age eligible for health insurance of old people	70+
cf. age elifible for Medicare in USA	65+
labor policy in Japan	
old population	65+
productive population	15-64
young population	0-14
aged worker	55+
middle-aged and aged worker	45+
minimum age for mandatory retirement	60
international policy	
aging rate	population of 65+/entire population
aging socitety	aging rate ≧ 7%
super-aging society	aging rate ≧ 25%
elderly household	household with 65+
survey statistics	
oldest category of census in USA	85+
oldest category of census in Japan	100+
young-old	65-74
middle-old	75-84
old-old, oldest-old	85+
very-old	80+

DISCUSSION

Most of the existing social security systems for aged people were developed in the days when retired people did not live long and it became adequately established during world economic expansion. Achievement of long and peaceful years after retirement has been the desired result of the system. Yet, the aging occurring at present is serious because it is occurring simultaneously with economic recession in many counties. It poses several essential questions to the policy makers in each society:

(1) whether the society should spend wealth saved by the older generation, even if it may cause economic damage to the future pension system (conservative policy),

(2) whether the society should reform the pension system so as not to require too much contribution from the younger generation, even if it may require unfair reduction of the once promised benefit for aged people (pension reform policy),

(3) whether the society should motivate aged people to transmit work opportunities to young people, even if it may burden unemployment insurance (employment adjustment policy),

(4) whether the society should protect the human right of aged people to work or resign, even if it may increase the unemployment rate among young people (human right protection policy),

(5) whether the society should arrange living and working environments more appropriate for aged people, even if it may not be efficient for productivity due to the extra cost (health and safety protection policy), and

(6) whether the society should prepare a social environment to encourage young people to bear and raise more children, even if it may cause an undesired increase in world population (birth promotion policy).

Most countries seem not to be able to afford continuation of a conservative policy, and simply await the outcome of the birth promotion policy. Pension reform and employment adjustment are the two practical and expected policies from their economic impact, and are adopted by many countries in diverse forms [6]. Human right protection policies in some countries abolished the words "mandatory retirement" or "early retirement" because they sound like forcing people to inactivity at a certain biological age. Alternatively, these countries often have special unemployment schemes with bonus benefit for aged people. Although it is hard to distinguish these two policies, the difference may lie in whether aged people wish and are capable to work. In these discussions on policy to cope with aging of society, it is often observed that the issue of health and safety protection is neglected in favor of the argument on economics. We could not find statements in legislation or policies that clearly dealt with the matter of improving the working environment exclusively for aged workers, except for the Japanese Law. We supposed it is probably because the involvement of the terminology related to biological age in these sentences may not be justifiable in many countries; however in Japan, this is not the case, possibly because words related to disability may be more misleading in the Japanese language. Extensive research and training are now being conducted to measure the physiological and mental effects of aging and to improve the working environment for aged people [7-10]. Appropriate preparations for accepting them in our societies can never be overemphasized.

To summarize, we are now looking for a compromise point of complicated interests relating to human rights, the economy, and health protection. The interrelation of these issues is summarized in Figure 4.1. Political decisions may depend much on the robustness of the economy, infrastructure to accept other shortcomings from the proposed decision, variation of accommodated choices that the policy makers can make, and cultural views of aged people. Noting that the terminology of aged people often imagines retired people, the definition of aged people may vary along with the change of actual retirement age derived from the final policy.

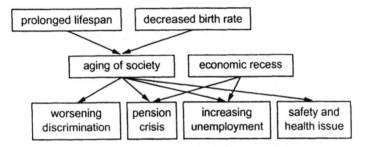

Figure 4.1 Interrelation of issues on aging

CONCLUSION

This study described the international similarities and differences of both the policies formulated for the aging society and the vocabulary associated with age. The extent of prolonged life span, low birth rate, economic recession and cultural background differently affect the local condition of aging. Policies combating the pension crisis, unemployment and age discrimination were interrelated. Mandatory retirement is legally prohibited in the USA, Canada and New Zealand, whereas in some other countries it is admitted as a rational system to share work opportunities with the younger generation. There are wide differences across countries regarding retirement age and pensionable age, and among the relevant policies. Extensive research has been performed relating to safety and health protection for aged workers; however, the findings were not often applied nor literally stated in legislation. The most widely prevailing definition of aged people implies people of age 65 or over. Related terms such as "aged," "old," "elderly," "young-old," "old-old," and "middle aged" were used with different terminology across the countries. Clear definitions were applied in policies of pension and insurance, but not necessarily in labor policies. The notion of "aged worker" was generally younger than "aged people." International standardization and cooperation would be a concern for the future. Applied research and technological challenges should be further pursued with regard to the objective measurement of functional performance of workers and the evaluation of work adaptation of aged workers, as well as the establishment of a balanced social system for human rights, the economy and health protection for the aging society.

REFERENCES

1. OECD,1998, Workforce ageing in OECD countries: employment outlook. *OECD*, pp 123-151
2. de Zwart B., Frings-Dresen M. and van Dijk F., 1995, Physicalworkload and the ageing worker: a review of the literature, *Int Arch Occup Environ Health* **68**, pp.1-12
3. Rabbitt P., 1991, Management of the working population, *Ergonomics 34*, pp. 775-790
4. Westerholm P. and Kilbom A., 1997, Aging and work: the occupational health services' perspective. *Occup Environ Med 54(11)*, pp. 777-780
5. Takegawa S. and Shionoya Y.,1999, Social security systems of the advanced countries vol 1- vol 6, (Tokyo: University of Tokyo Press)
6. OECD, 1998, Maintaining prosperity in an ageing society: the OECD study on the policy implications of ageing, *OECD*.
7. Ilmarinen J. and Tuomi K., 1991, Summary and recommendations of a project involving cross-sectional and follow-up studies on the aging worker in Finnish municipal occupations (1981-1985). *Scand J Work Environ Health 17(1)*, pp. 135-141
8. Rantanen J., 1999, Research challenges arising from changes in worklife, *Scand J Work Environ Health 25(6)*, pp. 473-483
9. Mangino M., 2000, The aging employee. Impact on occupational health. *AAOHN J 48(7)*, pp. 349-357
10. Davis P. and Dotson C., 1987, Job performance testing, an alternative to age discrimination. *Med Sci Sport Exer 19*, pp.179-185

5. Employment of the Elderly in Korea

Kwan S. Lee and Jae H. Kim

Department of Information Industrial Engineering,
Hong-ik University, Seoul, Korea

ABSTRACT
This paper presents results of a survey of the employment status of the elderly in Korea. The status of employment of elderly people older than fifty-five years was studied using the labor panel study conducted by the Korea Labor Research Institute in 1998. Income, working hours, job, job satisfaction, job search effort and work training for persons who were employed before age 50 and after age 50 were analyzed. It was found that the elderly had difficulty in getting jobs after they reached 50 years old. Most jobs that they could find were temporary jobs. Their income was reduced significantly and they were not satisfied with their work and income.

Keywords: *Elderly, Employment, Job, Income*

INTRODUCTION

Korea has become an aged society according to the UN classification. In Korea, as in many developed countries, the aged population will increase drastically and the number of single aged people also will rise. As the size of this aged population grows, support for the elderly becomes a very important social issue (Chung, K. H. and Oh, Y. h., 1998). Although Korea still maintains the traditional family style, a large proportion of the younger generation shows a tendency to neglect their responsibility of supporting their parents. Therefore, it is important for the elderly to find jobs if they do not have sufficient savings. However, it is not easy to find a job once they are old (Hwang, J. S., 1993). The economic problem of the elderly is becoming a social problem in Korea.

It is the objective of this study to find reasons for the difficulty experienced by the elderly in finding jobs in Korea.

METHOD

In this study, the status of employment of elderly people older than fifty-five years was studied using the labor panel study conducted by the Korea Labor Research Institute in 1998. In this survey, 2,527 persons were interviewed. Income, working hours, job, job satisfaction, job search effort and work training for persons who were employed before age 50 years and after age 50 years were analyzed.

RESULTS

1. Employment Status of Elderly Workers

Table 5.1 shows the status of employment of elderly workers who were surveyed. It was found that two-thirds of elderly persons were unemployed. Among them, 1,527 people (91.2%) did not want to have jobs. Therefore, the number of real unemployed persons was 148, which is 14.8% of the total number of those who wanted to find jobs. This figure could be misleading, because the survey was carried out when Korea was in the midst of economic difficulties. However, 14.8% is much higher than the normal unemployment rate for all Koreans at that time. Among employed elderly workers, only 39.2% were regular salaried workers, which is quite a bit lower than the rate for all adult workers.

Table 5.1 Employment status

(Unit: %)

Employment status	Number (%)	Before age 50	After age 50	Classification	Number (%)
Employed	852 (33.4)	530 (37.8)	322 (62.2)	Salaried workers	334 (39.2)
				Non-salaried workers	518 (60.8)
Unemployed	1,675 (65.6)			No jobs	148 (8.9)
				Retirement	1,527 (91.2)
Total	2,527 (100.0)				

2. Income, Type of Jobs, and Union Status of Elderly Workers

Table 5.2 shows incomes of the elderly who were employed. It was found that their income decreased by 25% after they reached age 50. There is not much difference in income regardless of their employment time. Their average income was a little less than $600 per month.

Table 5.2 Monthly income

(Unit: %)

Amount	After age 50		Before age 50
	Present	Past	
Less than 500,000 Won	35.2	32.9	48.4
500,000 - 1,000,000 Won	44.2	30.3	21.0
1000,000 - 1,500,000 Won	9.3	16.9	13.8
1,500,000 - 2,000,000 Won	3.3	6.1	6.2
Over 2,000,000 Won	8.0	13.9	10.6
Mean	72.4 (100.0)	96.5 (100.0)	73.2 (100.0)

Table 5.3 shows jobs of elderly workers. It shows a shift in employment from more professional jobs and jobs requiring skill and knowledge to simple sedentary jobs if elderly people look for jobs after they reach age 50. Elderly workers who kept their former jobs were mostly farmers and fishermen, followed by service workers.

Table 5.3 Type of jobs

(Unit: %)

Type of jobs	After age 50		Before age 50
	Present	Past	
Government workers, Managers	2.2	1.9	2.4
Professionals	2.2	1.9	4.5
Engineers	8.8	12.5	5.0
Sedentary workers	3.8	6.8	2.2
Service workers	19.7	20.2	19.9
Farmers and Fishermen	14.1	10.3	50.0
Operators	4.7	11.0	5.8
Assembly workers	5.9	11.0	2.8
Laborers	38.8	24.3	7.4

Table 5.4 shows that many elderly workers ware hired as non-salaried workers if they looked for jobs after age 50.

Table 5.4 Type of payment

(Unit: %)

	After age 50	Before age 50
Salaried workers	63.1	83.5
Non-salaried workers	36.9	16.5

Table 5.5 shows that the working hours were longer if people found jobs after age 50, and most of them worked longer than 45 hours per week. It was found that elderly workers had less opportunity (63.1% vs. 83.5%) to join a union if they were hired after age 50. This may be partly due to the fact that many elderly workers were non-salaried workers who were not qualified to be union workers (75.0%).

Table 5.5 Working hours

(Unit: %)

	After age 50	Before age 50
0 - 31 hours	15.5	16.9
32 - 44 hours	10.2	12.7
Over 45 hours	74.3	70.4

3. Job Satisfaction and Life-related Satisfaction

Table 5.6 shows the job satisfaction of these elderly workers. More than 50% of elderly workers were not satisfied with their jobs, regardless of when they were employed.

Table 5.6 Job satisfaction (before or after age 50)

(Unit: %)

	Income		Job security		Job		Personal progress		Work condition	
	before	after	before	after	before	after	before	after	before	after
Satisfied	12.1	8.7	24.4	18.5	30.4	25.3	12.7	10.7	23.1	21.2
Acceptable	28.0	28.3	41.2	38.8	41.6	45.8	47.5	48.4	43.2	48.3
Dissatisfied	60.0	62.9	34.5	42.9	28.1	28.9	39.8	40.8	33.7	30.5

Table 5.7 shows the daily life satisfaction of elderly workers in five categories: life overall, household income, family relationships, leisure activities and residential situation. It shows that they indicate satisfaction with family relationships and residential situation. However, family income and leisure activities do not satisfy them in general. This result is understandable, since they work partly because of income reasons and thus they do not have enough time for leisure.

Table 5.7 Daily life satisfaction

(Unit: %)

	Life overall		Household income		Family relationships		Leisure		Residential situation	
	before	after	before	after	before	after	before	after	before	after
Satisfied	20.4	17.7	13.4	10.8	63.8	53.3	23.3	15.8	40.2	32.2
Acceptable	46.1	41.3	33.3	29.0	30.5	38.8	40.3	46.7	45.2	46.7
Dissatisfied	33.6	41.0	53.2	60.3	5.7	7.9	36.4	37.6	14.5	21.2

4. Job Search

Table 5.8 shows types of difficulties which elderly workers encountered when they looked for jobs. Almost all of them, regardless of their employment time, stated that there were not enough jobs. They found that age was the leading reason for difficulty in getting jobs if they tried to find jobs after age 50, while the lack of jobs was the leading reason when they tried to find jobs before age 50. Almost all of them also complained of lack of information about job openings. Interestingly, fewer people complained about lack of information if they were hired after age 50. This may be due to the fact that many elderly workers are looking for simple sedentary jobs which are mostly advertised in the newspaper. It is also interesting to discover that they experienced lack of qualifications, lack of experience, lower income, inappropriate working conditions, and age discrimination when they looked for jobs before they turned 50, but they experienced these problems less when they tried to find jobs after they turned 50. Caution may be required in interpreting this fact. It may be due to the fact they are not as demanding or expecting of jobs as were other elderly people when they looked for jobs before age 50.

Table 5.8 Types of difficulties in finding jobs

(Unit: %)

	Lack job		Lack of information		Lack of a qualification		Lack of experience		Lower income		Working condition / hours		Age	
	before	after	before	after	before	after	before	after	before	after	before	after	before	after
Strongly agree	66.7	54.5	33.3	27.3	33.3	9.1	66.7	27.3	-	18.2	-	9.1	33.1	36.4
Agree	33.3	36.4	66.7	54.5	66.7	45.5	33.3	54.5	100.0	36.4	66.7	27.3	66.7	45.5
Disagree	-	9.1	-	18.2	-	45.5	-	18.2	-	36.4	33.3	36.4	-	18.2
Strongly disagree	-	-	-	-	-	-	-	-	-	9.1	-	27.3	-	-

5. Work Training

Table 5.9 shows that most of them (8%) did not have work training experience before, and they did not want to have new work training. This may reflect their low expectations of the new job.

Table 5.9 Types of difficulties in finding jobs

(Unit: %)

	After age 50	Before age 50
Have experienced	5.6	8.0
Experiencing now	0.3	-
No experience	94.1	92.0

CONCLUSIONS

It was found that the elderly had difficulty in getting jobs after they reached age 50. Most jobs that they could find were temporary jobs. Their income was reduced significantly and they were not satisfied with their work and income.

REFERENCES

1. Chung, K.H. and Oh, Y.H., 1998, A survey of elderly life and welfare requirement, Korean Institute for Health and Sociology
2. Hwang, J.S., 1993, A search for promotion for elderly employment, *Korean Gerontology Vol. 13*, **No 1**, pp.158.

PART II
Measures for Healthy Aging: Lifestyle, and Exercise

6. Lifespan Functional Fitness: Encouraging Human Struggle (Physical Activity) and Warning About the Cost of Technology

Max Vercruyssen[1,2]

[1] *UNIVERSITY OF HAWAI'I, John A Burns School of Medicine, Geriatric Medicine Program, Gerontechnology & Neuromuscular Research; John A Hartford Center for Excellence in Geriatrics; Pacific Islands Geriatrics Education Center, Honolulu, Hawaii USA*
[2] *HAWAI'I ACADEMY - a private school for the advancement of lifetime fitness, human sciences, and technology, Honolulu, Hawaii USA*

ABSTRACT

In recent decades, health, life expectancy, work conditions, and the quality of later life have improved considerably in nearly all industrialized countries. However, advances in technology, particularly rapid development of labor saving devices, the shift from manufacturing to information management, and attitude changes caused by reductions in life demands has fueled an epidemic of inactivity, obesity, and sometimes apathy toward social and physical activities. Current sedentary living and attitudes about work efforts are a serious threat to the quality of current and future work productivity and living conditions. This paper (1) describes some of the unexpected costs of technology and warns of an epidemic of inactivity that threatens the quality of present and future work and retirement life; (2) introduces the term 'lifespan functional fitness;' and (3) proposes that one's attitude, particularly the struggle in addressing the challenges of aging, may be the most important variable in the wellness and productivity of senior and retired workers.

Keywords: *Lifespan functional fitness, Human aging, Elderly workers, Geriatrics, Work ability, Fitness, Health, Wellness, Independent living, Functional autonomy, Retirement, Gerontechnology, Ergonomics, Human factors, Technology, Activities of daily living, Strength, Flexibility, Balance, Agility, Power, Inactivity, Obesity, Epidemic, Life expectancy*

INTRODUCTION

Early Workers. In the Stone Age (prehistoric times), life expectancy was about 18 years and death was usually caused by an occupational accident - while hunting or gathering food, the worker became the meal for a dinosaur or other predator. In ancient Greek and Roman times the life expectancy was 20-22 years and most deaths were from battle, epidemic disease, and accidents. In the 1600s, humans were living to about 35 years of age. In 1900, the life expectancy for Americans was 47 years, making the American Social Security (retirement and life insurance) System a financial winner for the government because it was based on continual pay-in until a pay-out (retirement) beginning at 65 years of age (most people never collected). The Industrial Age was characterized by improved overall health and working conditions, a reduced number of work-related deaths, and a life expectancy of nearly 70 years.

Present Day Workers. Industrialized countries are now in the Information Age, a period of health and prosperity, relatively free of diseases, war, and natural disasters. We have more leisure time than ever before, much due to numerous labor saving and information management devices. The current life expectancy is approaching 80 years, less for males. Workers today are inclined to spend most of their time in sedentary activities and then spend their leisure time the same way ... usually watching monitors that display information for entertainment. Excluding genetics, the leading cause of all deaths in the United States in 1990 was smoking and the second largest underlying cause of death was a lack of exercise and poor dietary habits [1]. Unfortunately, the epidemic of inactivity among children and working adults is impacting health systems and the wellness of everyone in ways we are just beginning to understand. Society has been caught off-guard by the large number of individuals that live past retirement longer than before and whose roles are undefined or still evolving. Technology may help us compensate for and prevent some of the debilitating effects of aging, but this chapter argues that it is the 'struggle,' i.e., the perseverance and steadfast attempt at coping to remain functional, that enhances longevity and the quality of one's later life.

The purpose of this chapter is to (1) describe some of the unexpected costs of technology and warn of an epidemic of inactivity that threatens the quality of present and future work and retirement life; (2) introduce the term 'lifespan functional fitness;' and (3) propose that one's attitude, and struggle in addressing the challenges of aging, may be the most important variable in the wellness and productivity of senior and retired workers.

UNEXPECTED COSTS OF TECHNOLOGY

Technology has changed the very way we live. Its social impact is much greater than we can imagine. However, there are many hidden costs of technology. This section presents some of those costs, including acquired dependence on technology, frequent over-use of technology, and some lost opportunities caused by the misuse of technology. The ineffective use of technology is a primary cause of our current epidemic of inactivity (and obesity) and is directly responsible for excessive health care costs that will certainly catch society off-guard.

Acquired Dependence on Technology. Everyone uses some form of technology and would be inconvenienced without it (e.g., take away electricity and notice how everyday activity becomes much more difficult). Whether we want it or not, technology is usually forced upon us and then we become dependent on it and have difficulty functioning without it. For instance, automatic teller machines at banks are typically made available free and it takes some doing to get customers to use them (incentives, after-hours convenience, etc.). Eventually customers can be charged for their use because they have become comfortable with the process. Repayment of the initial investment for ATMs comes when the bank has increased customer volume, reduced human teller costs, and customer charges to do business with a human. Whatever the item, regardless of the method of introduction, we become dependent on highly used technologies.

Frequent Overuse of Technology. We have strived to make life more convenient by inventing labor saving devices and we have become quite good at it. Even the smallest of physical and mental activities we try to reduce or eliminate. For instance, changing stations on a television. Once seated in a living room it has become commonplace for remote devices to keep the consumer from getting out of their chair. Often problems emerge when the remote control is not readily available. Many children are growing up unable or unwilling to operate the television manually. Another example is the portable calculator: it is sad that many people today cannot or refuse to try to do simple arithmetic without this tool. Dependence on a technology is directly linked to its overuse. Using an elevator instead of walking up one or two flights of stairs is a convenience ... if that person performs other physical activities that help maintain health and fitness, otherwise convenience of the elevator will take its toll later in a reduced quality of later life.

Humans need a certain amount of physical activity to maintain health, fitness, and wellness. Labor saving systems are wonderful, especially if they reduce risk of or exposure to occupational hazards; however, if use of such systems causes insufficient activity for optimal well-being, operators should be warned of their overuse. Walking, swimming, and even cycling can be viewed as manual modes of transportation and have great health benefits from regular use. Overuse of automatic modes, like cars, trains, and planes, poses a public health problem that is yet unaddressed.

Opportunities Lost by Use of Technology. Today, much of our work and leisure time is spent watching a variety of different monitors and displays in relative isolation from others. We often sit fixated on information displayed while ignoring others around us. Unless there is physical activity elsewhere in our life, this sedentary living is very costly in terms of physical, mental, and social health, as well as other opportunities lost. We have become families of 'watchers' instead of 'doers.' In the past, children played hard, outdoors, as long as they could before having to come indoors (often totally exhausted) for meals or other family activities. Now, especially in urban areas, children are transported to school, where they get little or no physical activity, and then transported home where they stay indoors, often with their only outside contacts coming from the internet. The information age has brought opportunities to enhance knowledge and intellectual growth but sometimes it is at a cost to physical, mental, social, emotional, and vocational development. In other words, surfing the internet has value but prevents the surfer from interacting with others, reading a book, and developing other psychomotor and socializing skills. The situation is even worse for adults.

Epidemic of Inactivity Caused by Misuse of Technology. The number of inactive individuals in our society is growing at an alarming rate. Children, teenagers, and adults have become sedentary and risk health complications (especially mobility problems) directly due to their inactivity [e.g., 2]. Among those over 65 years in the United States, only about 25% of men and 20% of women meet the national guidelines for physical activity (most are totally sedentary) and nearly 60% of this population fails to obtain sufficient amounts of regular activity needed to reduce their risk of premature morbidity and mortality [3]. Further, inactivity increases with age such that by age 75, about one in three men and one in two women engage in NO physical activity. By the age of 80, four out of five people report some disability, of which cardiovascular disease and arthritis are the most common reported. Our quest for convenience and labor savings has led us to our current sedentary lifestyle.

Epidemic of Obesity Caused by Inactivity. Not surprisingly, those countries that are observing increases in inactivity are also experiencing an obesity epidemic. In the United States, the prevalence of obesity (body mass index $\geq 30\text{kg/m}^2$) increased from 12.0% in 1991 to 17.9% in 1998 with a steady increase observed in all states; in both sexes; across age groups, races, educational levels; and occurred regardless of smoking status [4; see also 5]. The greatest magnitude of increase was found in 18-29 year olds (7.1% to 12.1%), those with some college education (10.6% to 17.8%), and those of Hispanic ethnicity (11.6% to 20.8%). Researchers at the University of Hawaii recently have found evidence that Type 2 diabetes - often the result of obesity and a sedentary lifestyle - is appearing at ever younger ages because of a generation raised on television and fast food. It is no longer occurring primarily in adults 60 years and older; many children today are at risk of having major cardiovascular disease or other complications in their 30s or 40s (Honolulu Heart Study; Pacific Health Research Institute).

Warning: Anticipate a High Cost of Physical Inactivity. In a cross-sectional stratified analysis of the 1987 National Medical Expenditures Survey that included US civilian men and non-pregnant women aged 15 and older who were without physical limitations and not residing in institutions, those who engaged in regular physical activity had direct medical costs of $1,019 compared with $1,349 for those who were inactive. The costs were lower for active compared to non-active persons among smokers ($1,079 vs. $1,448) and nonsmokers ($953 vs. $1,234) and were consistent across age-groups and by sex. Medical care use (hospitalizations, physician visits, and medications) was also lower for the physically active people than for inactive people. In 1993 dollars, the medical costs of physical inactivity ($45.6 billion) appear to be similar to those due to smoking ($50.0 - 53.4 billion) [6]. The authors speculate that increasing participation in regular moderate physical activity among the more than 88 million inactive Americans over 15 years might reduce annual medical costs by as much as $29.2 billion in 1987 dollars - $76.6 billion in year 2000 dollars. [6; see also 7].

Assuming the pattern of direct and indirect costs attributable to physical inactivity is similar to those for coronary heart disease, diabetes, and obesity [e.g., 8], Pratt et al. [6] estimate the total direct and indirect costs associated with physical inactivity may well be in excess of $150 billion in 2000 dollars. Obesity-related morbidity alone may account for 6.8% of all US health care costs [9]. We are only just beginning to notice the enormous economic impact of inactivity and obesity [see also 10].

> **Society needs to shift from a sedentary to an active lifestyle ...**
> **our health and quality of life depends on this change!**

LIFESPAN FUNCTIONAL FITNESS

Lifespan Functional Fitness is simply the *ability to do what one wants to do,* throughout one's lifetime. *Lifespan* is the period between birth and death. *Functional* refers to the functions of daily living, including self-care, home chores, and the tasks, jobs, duties, and goals found in work, school, and leisure activities. *Fitness* describes the capacity to meet regular and emergency task demands, including such concepts as capacity needed for the task, total capacity for immediate responses, and reserve capacity (reserve = total - needed capacity; important for determining stress and recovery in the workplace and in private life). Functional abilities and functional limitations are similar concepts [e.g., 11]. Work fitness and fitness for work/duty, has been described in the work-ability index [e.g., 12-14].

Physical fitness involves the ability to independently perform daily tasks vigorously and alertly, with energy left over for enjoying leisure-time activities and meeting emergency demands. It is the ability to endure, to bear up, to withstand stress, to carry on in circumstances where an unfit person could not continue, and it is a major basis for good health and well-being. It is a condition that helps us look, feel, and do our best.

Lifespan Functional Fitness concepts and principles apply to all ages but this discussion focuses briefly on older adults in the workforce and retirees. Senior worker wellness, productivity, and quality of life varies within and among individuals as a function of at least three dimensions for each task or function to be performed: (1) **Age/Cohort Continuum** (Infant needing care, Child in school, Adult in workplace, Senior in retirement, Elderly needing care); (2) **Ability/Output Continuum** (within each age group the highest functioning can be called elite to those requiring assistance, the frail); and (3) **Intervention Components** (including categories for Physiological, Psychological, Sociological, and Assistive Technologies).

Kumashiro [15] has advanced ergonomics strategies for optimizing productivity with older workers. Viewing productivity as a function of the interaction of five variables (functional age, health condition, ability and work capacity, environment / work conditions, and motivation). Kumashiro's ERGOMA approach is a model into which lifespan functional fitness could be inserted to realize optimal work outputs in senior workers. Likewise, functional fitness has a place in Ilmarinen's [16,17] models of the factors influencing work ability, including health and functional capacities, education and competence, motivation and job satisfaction, values and attitudes, and work and community environment. His research group has also developed strategies for promoting work ability [18] and the concept that work ability is the intersection of overlapping spheres for true biological aging, lifestyle, work, and health [19].

One's fitness to perform required activities of daily living, community tasks, or obligations is a useful index for predicting work ability and the stress experienced by the worker (the effect on reserve capacity). Further, functional fitness tests that have been developed through research to be valid and reliable (e.g., the Senior Fitness Test [20,21]) can be used to profile the individual to indicate their strengths and weaknesses, identify those at risk of losing functional ability, and prescribe rehabilitation exercises. Presumably, physical frailty commonly associated with aging could be prevented if weaknesses were detected and treated before they become overt losses of functional ability. Currently, in America, about 10% of individuals between 65 and 75 years of age need assistance with their daily activities; for those between 75 and 85, the percentage increases to about 25%; and by 85, about 50 percent require assistance. Clearly, 'the benefits associated with regular exercise and physical activity contribute to a more healthy, independent lifestyle, greatly improving the functional capacity and quality of life in this population' [22, p. 992].

Ultimately, one's individual happiness and well-being is an end result of the interaction (and balance) of six dimensions or domains of wellness (Physical, Intellectual, Social, Emotional, Spiritual, and Vocational) and the innate disposition of the individual with at least six external conditions (Environment, Physical, Social, Economic, Political, and Exercise). According to the positive health life-style [23] and the whole-istic approaches to wellness [24], the **physical dimension** focuses on maintaining independence by development of the ability to exercise properly, to eat a healthy diet, to avoid high-risk behaviors, and to make positive lifestyle choices. The **intellectual dimension** involves the ability to increase knowledge, think critically, to identify and solve problems, and to use information to enhance personal development - to increase creativity and gain a better understanding and appreciation for oneself and others. The **social dimension** includes development of meaningful personal relationships (and harmony) with family and friends. The **emotional dimension** focuses on the development of self-confidence and a positive self-concept, the ability to handle stress, the ability to express emotions appropriately, and to accept one's own limitations. The spiritual dimension involves the ability to find meaning and purpose in life, to develop faith in nature, religion, or a higher entity to develop values and enhance moral and ethical development. The **vocational dimension** is concerned with establishing and achieving personal interests and growth through meaningful activities like volunteering in the community.

THE STRUGGLE FOR FUNCTIONAL FITNESS

Life is a struggle ... and it gets worse as we get older. A strong spirit is needed to confront the challenges of daily living and to persevere when things seem overwhelming. In some cases, successful aging may be viewed as simply an attitude, an expression of the drive to conquer whatever obstacles are thrown in the way of your goals. To struggle is to live ... give up and we die. Promoting the struggle for one's functional fitness positively impacts communities of all ages and abilities, and fosters improvements in work productivity and the quality of work life, especially for seniors in the work force. Courage is contagious!

Barriers in the Workplace. Complex barriers impede efforts to increase and maintain physical activity among older adults in the workplace. According to the National Blueprint (document in development), good economic models are needed to illustrate cost effectiveness to employers of increasing physical activity among older employees. Employers may have concerns about liability and worry about implications of employees being injured or becoming ill while participating in on-site physical activity programs or events. There is little information available that workplace exercise programs are effective and what measurable outcomes are most persuasive to management (e.g., are employers most interested in improved performance and productivity, reduced health care costs, or reduced absenteeism?). Workplaces have a high degree of variability, which may make on-site physical activity programs of older workers challenging to implement. [25]

Addressing Barriers and Setting Strategic Directions in the Workplace. In the United States, a national plan (blueprint; 25) for increasing physical activity in the age 50 and older population is about to be released. It recommends strategies based on the assumption that people generally work in or near the community in which they live, and work sites can often operate as a community resource or center to: (1) Seek employee input in the planning and development of programs targeted to the age 50 and older worker. (2) Create a workplace environment where time for physical activity is incorporated into daily activities. (3) Design a system that provides employers with tax incentives based on physical activity programs/opportunities they afford their employees. (4) Provide financial incentives to employers that incorporate physical activity enhancements in their corporate setting land-use plans. (5) Provide health insurance cost reductions to employers that offer physical activity programs to employees. (6) Develop, implement, and evaluate model work site physical activity programs, targeted to employees age 50 and older. (7) Provide tools and templates to enable employers to communicate information about the importance of physical activity. (8) Identify and disseminate information about successful work site physical activity programs designed for employees age 50 and older. (9) Communicate to business leaders the benefits of physical activity for older workers, especially as they pertain to desired outcomes of management (e.g., cost-savings, employee absenteeism, etc.). [25]

Examples of Courageous Struggles. One promising intervention or marketing strategy is to highlight examples of individuals who have overcome great obstacles to remain relatively healthy, happy, and functionally independent. These examples motivate everyone, regardless of age or ability. Spotlighting our community heroes stimulates others to struggle harder.

> Rekindle That Part of the Human Spirit That Drives One to
> Struggle Against All Odds, To Persevere With Courage and
> Determination, To Inspire Others Not to Give Up Without a
> Fight...Bushido!

CONCLUSION

- Anticipate a demographic explosion. Populations are graying throughout the world - countries are graying and the most rapidly growing segment is the 85+ years cohort. Emphasize approaches and designs with a lifespan and Transgenerational perspective.
- Technology is NOT a cure-all for improving living conditions, in fact, frequent use of labor-saving (assistive or adaptive) devices and a sedentary lifestyle may be injurious to one's wellness and productivity.
- The spreading **epidemic of physical inactivity** (and obesity) will cost society greatly in terms of direct and indirect health care - probably in excess of 150 billion US dollars this year and in America and increasing exponentially each year.
- **Lifespan Functional Fitness** is a useful concept for encouraging an intervention to offset the debilitating effects of physical inactivity and to promote personal productivity in later life, especially for seniors in the workforce.
- Emphasize the effective use of technology, especially with respect to using high-technology, low-technology, and 'no-technology' solutions and interventions.
- Ultimately, one's attitude about aging and exercise is likely the single most important variable for insuring optimal quality of later life ... **struggle hard, ... live well, ... be happy!**

ACKNOWLEDGMENTS

This research is part of an ongoing effort concerned with Lifespan Functional Fitness and is funded in part by (1) a grant from the University of Hawaii and Kapiolani Clinical Research Center (CRC 98-18); (2) support from the University of Hawaii Geriatric Medicine Program, John A Burns School of Medicine, the Pacific Geriatric Education Center, and the John A Hartford Center for Excellence in Geriatric Medicine; (3) funding from Hawaii Academy's Research and Clinical Services Departments; (4) a stipend from the University of Occupational and Environmental Health for presentation in the 21st UOEH and 4th IIES International Symposium; and (5) a stipend from the Georgia Institute of Technology, Center on Aging and Cognition (Edward R. Roybal Center for Research on Applied Gerontology; National Institute on Aging) for presentation at the 2000 Conference on Human Factors and Health Care Interventions for Older Adults.

REFERENCES

1. NIA, 1998, *Exercise: A guide from the National Institute on Aging*, Publication No. NIH 98-4258 (free on request), 1.800.222.2225, (Public Information Office: National Institute on Aging, National Institutes of Health, Bldg 31, Rm 5C27, 31 Center Drive, MSC 2292, Bethesda, MD 20892-2292).
2. Buckwalter, J.A., 1997, Decreased mobility in the elderly: The exercise antidote. In *The Physician and Sportsmedicine*, **25**(9), pp. 127-186.
3. U.S. Department of Health and Human Services, 1996, *Physical Activity and Health: A Report of the Surgeon General*, pp. 85-172, (Atlanta, GA: U.S. Department of Health and Human Services, Centers of Disease Control and Prevention, National Center for Chronic Disease Prevention and Health Promotion).
4. Mokdad, A.H., Serdula, M.K., Dietz, W.H., Bowman, B.A., Marks, J.S. and Koplan, J.P., 1999, The spread of the obesity epidemic in the United States, 1991-1998. In *Journal of the American Medical Association*, **282**(16), pp. 1519-1522.
5. Flegal, K.M., Carrol, M.D., Kuczmarski, R.J. and Johnson, C.L., 1998, Overweight and obesity trends in the United States: Prevalence and trends, 1960-1994. In *International Journal of Obesity and Related Metabolism Disorders*, **22**, pp. 39-47.
6. Pratt, M., Macera, C.A. and Wang, G., 2000, Higher direct medical costs associated with physical inactivity. In *The Physician and Sportsmedicine*, **28**(10), pp. 63-70.
7. Nicholl, J.P., Coleman, P. and Brazier, J.E., 1994, Health and healthcare costs and benefits of exercise. In *Pharmoeconomics*, **5**(2), pp. 109-122.
8. Wolf, A.M. and Colditz, G.A., 1998, Current estimates of the economic cost of obesity in the United States. In *Obesity Research*, **6**(2), pp. 97-106.
9. Wolf, A.M. and Colditz, G.A., 1996, Social and economic effects of body weight in the United States. In *American Journal of Clinical Nutrition*, **63**(suppl 3), pp. 466S-469S.
10. Colditz, G.A., 1999, Economic costs of obesity and inactivity. In *Medicine and Science in Sports and Exercise*, **31**(1), pp. S663-S667.
11. Morgan, L. and Kunkel, S., 2000, Chapter 10: Health and health care. In *Aging: The social context, 2nd.* (New York: Pine Forge).
12. Ilmarinen, J., 1995, Aging and work: The role of ergonomics for maintaining work ability during aging. In *Advances in industrial ergonomics and safety VII*, edited by Bittner, A.C. and Champney, P.C., (London: Taylor & Francis), pp. 3-17.
13. Ilmarinen, J., 1999, *Ageing workers in the European Union-Status and promotion of work ability, employability and employment*, Finnish Institute of Occupational Health, Publications Office, Topeliuksenkatu 41 a A, FIN-00250 Helsinki (FIM 190).
14. Ilmarinen, J. and Louhevaara, V., 1999, FinnAge-Respect for the ageing: Action programme to promote health, work ability and well-being of aging workers in 1990-1996. In *People and work, Research Report 26*, Finnish Institute of Occupational Health, Publications Office, Topeliuksenkatu 41 a A, FIN-00250 Helsinki (FIM 160).

15. Kumashiro, M., 2000, Ergonomics strategies and actions for achieving productive use of an ageing work-force. In *Ergonomics*, **43**(7), pp. 1007-1018.
16. Ilmarinen, J., 2001a, Aging workers. In *Occupational and Environmental Medicine*, **58** (8), pp. 546-552.
17. Ilmarinen, J., 2001b, Functional capacities and work ability as predictors of good 3rd age. In *Physical fitness and health promotion in active aging*, edited by Shiraki, K., Sagawa, S., and Yousef, M.K., (Leiden, The Netherlands: Backhuys), pp. 61-80.
18. Ilmarinen, J. and Rantanen, J., 1999, Promotion of work ability during ageing. In *American Journal of Industrial Medicine Supplement*, **1**, pp. 21-23.
19. Ilmarinen, J., Huuhtanen, P. and Louhevaara, V., 1999, Developing and testing models and concepts to promote work ability during ageing. FinnAge-Respect for the ageing: Action programme to promote health, work ability and well-being of aging workers in 1990-1996, *People and work, Research Report 26*, edited by Ilmarinen, J. and Louhevaara, V., pp. 263-267, Finnish Institute of Occupational Health, Publications Office, Topeliuksenkatu 41 a A, FIN-00250 Helsinki (FIM 160).
20. Rikli, R.E., 2000, Evaluation of functional fitness in older men and women. In *Abstracts of the 20th International Symposium of the University of Occupational and Environmental Health Physiological Evaluation of Working Capability in Aged Laborers*, UOEH, Japan, pp. 15-16.
21. Osness, W.H., Adrian, M., Clark, B., Hoeger, W., Raab, D., and Wiswell, R., 1996, In *Functional fitness assessment for adults over 60 years (A field based assessment)*, 2nd, (Dubuque, IA: Kendall/Hunt). Developed by the Council on Aging & Adult Development of the American Association for Active Lifestyles & Fitness, Association of the American Alliance for Health, Physical Education, Recreation, and Dance.
22. Mazzeo, R.S., Cavanagh, P., Evans, W.J., Flatorone, M., Hagberg, J., McAuley, E., and Startzell, J., 1998, American College of Sports Medicine Position Stand on Exercise and Physical Activity for Older Adults. In *Medicine and Science in Sports and Exercise*, **30**(6), pp. 992-1008.
23. Williams, M.H., 1996, Lifetime fitness and wellness, 4th, *A personal choice*. (Chicago: Brown & Benchmark).
24. Armbruster, B. and Gladwin, L.A., 2001, More than fitness for older adults: A "whole-istic" approach to wellness. In *ACSM's Health and Fitness Journal*, **5**(2), pp. 6-28.
25. *National Blueprint: Increasing physical activity among adults age 50 and older*. (document in development), Sponsored by American Association for Retired Persons, American College of Sports Medicine, American Geriatrics Society, The Centers for Disease Control and Prevention, The National Institute on Aging, and The Robert Wood Johnson Foundation. (Princeton, NJ: The Robert Wood Johnson Foundation).

7. Health Status and Lifestyles of Elderly Japanese Workers

Takashi Muto[1], Hidehiro Sugisawa[2], Hye-kyung Kim[2],
Erika Kobayashi[2], Taro Fukaya[2], Yoko Sugihara[2], Hiroshi Shibata[3]

[1] Juntendo University School of Medicine, Tokyo, Japan
[2] Tokyo Metropolitan Institute of Gerontology, Tokyo, Japan
[3] Obirin University, Tokyo, Japan

ABSTRACT

This study was conducted to clarify the health status and lifestyles of elderly workers who held jobs following retirement at age 60. The subjects included 488 men between the ages of 61 and 64, selected from the results of a national survey conducted in 1999, with a sample size of 6,000 randomly selected Japanese. Structured interviews were performed, and health status and lifestyles were compared between workers (n=258) and non-workers (n=230) at the time of survey. Workers showed better subjective health status than non-workers. The proportion of workers and non-workers seen by doctors regularly was 44% and 52%, respectively, showing no significant difference. The mental health status was found to be better among workers than among non-workers. The number of undesirable lifestyles was smaller among workers than among non-workers, but there was no significant difference in the rate of the elderly having an undesirable lifestyle in 5 out of 7 lifestyles between the two groups. Although health status and lifestyles of elderly workers are generally superior to those of non-workers, a fairly large proportion of the elderly workers have unhealthy lifestyles and engage in jobs while having diseases. This indicates that health promotion programs targeting elderly workers should be implemented as one of the occupational health measures.

Keywords: *Aging, Elderly worker, Occupational health, Health status, Health promotion, Lifestyle, Japan*

INTRODUCTION

It has recently been claimed that Japanese society has entered a stage of "few births and deaths, with an aging population" [1]. The total fertility rate (aggregate total birth rate of women by age in a given year) has been on the decline, and has continued to record all-time low numbers, falling from 2.13 in 1970 to 1.34 in 1999 [2]. The average life span for Japanese males and females was 69.3 years and 74.7 years, respectively, in 1970. By 2000, however, it was 77.6 years and 84.6 years [2]. The population aged

65 or over in Japan was 7.4 million in 1995, accounting for 7.1% of the total population, and it is 22.3 million in 2000, accounting for 17.5% of the total population [2]. With this aging of Japanese society, it is feared that social security benefit expenditures such as medical care and pensions will increase tremendously [1].

Under these circumstances, the elderly are expected to be productive, rather than a burden on society [3]. With regard to the elderly themselves, 78% of those aged 60 or over believe that the retirement age should be 65 or over, and 93% of white-collar workers claim to be willing to continue to work past the age of 65 [4]. Reflecting these opinions, workers aged 60 or over represented 13% of the total work force in 1995, and this figure is estimated to be 20% in 2010 [5]. Therefore, the assurance of jobs for older workers is an extremely important issue in satisfying the needs of workers and maintaining vitality in the Japanese economy in the face of the aging of the population.

In order for elderly workers to be able to perform their jobs safely and with high productivity, their physical and mental health must be sufficient to fulfill their duties. Due to the lower health status of elderly workers, health care for the elderly has become increasingly important as an occupational health issue in Japan. One of the main reasons for this is that employers are responsible for workers' health and safety, so they may be held accountable if the health status of elderly workers with diseases or disorders is compromised by the performance of their jobs. Another reason is that most lifestyle-related diseases, once not a target of occupational health services, are now considered work-related diseases that employers are obliged to prevent [6].

The Ministry of Labor implemented a workplace health promotion scheme in 1988 called the "Total Health Promotion Plan," based on the Industrial Safety and Health Law, which stipulates that employers should make efforts to implement workplace health promotion. This policy has been implemented in response to the increase in lifestyle-related diseases among workers due to the increase in the number of middle-aged workers in their 40s and 50s, in order to cope with the possible increase in the number of accidents and injuries among them [7]. This program is intended to promote the physical and mental health of all employees, but its main target is middle-aged workers. The same type of health promotion programs should be implemented for elderly workers.

An effective health promotion program for elderly workers should be based on their needs, which are greatly influenced by their health status and lifestyles. The health status and lifestyles of elderly workers are expected to be better than those of the elderly without jobs, but they have not been fully clarified due to a lack of studies with a high-validity study design. Previous studies on this theme either had no comparison groups [8-10] or were conducted for subjects by non-random sampling and are not considered to be a representative sample of all Japanese elderly workers [10].

This study was conducted to clarify the health status and lifestyles of elderly workers who held jobs following retirement at age 60.

METHODS

The subjects included 488 men between the ages of 61 and 64, who had retired from their jobs at age 60. They were selected based on the data of a national survey conducted in 1999, which had a sample size of 6,000 (4,000 men and 2,000 women) randomly selected Japanese between the ages of 55 and 64. This survey was originally designed to investigate the retirement process of middle-aged and older workers in terms of their health and socio-economic status. Women were excluded from this study, as the number of women who retired from jobs at age 60 was small.

A cross-sectional design was employed in this study. The subjects were classified into one of two groups, workers and non-workers, at the time of the survey. Health status and lifestyles were compared between the two groups.

As indicators of health status, the following were used: distribution of self-rated health [11]; number and prevalence of subjective symptoms; rate of the elderly having difficulties with their vision, hearing, or everyday life; rate of the elderly seen by a physician at outpatient clinics; and mental health status. The following were employed as indicators of mental health status: mean of the Center for Epidemiologic Studies Depression (CES-D) scale; proportion of the elderly with 16 points or over on the CES-D scale [12]; and rate of users of psychotropic drugs such as sleeping pills, sedatives, or depressants. Six health practices, (smoking, drinking, breakfast, physical exercise, sleeping hours, and obesity) (but excluding food intake between meals from Breslow's well-known seven health practices [13]), plus annual health checkups, were used as indicators of personal lifestyle.

A personal interview with the subject was conducted by professional interviewers at the subject's residence. The interview was a structured one, using a questionnaire developed by the authors. The first interview was conducted from March to April 1999, and the second was conducted three months after the first, from June to July 1999, for those people for whom the first interview had not been conducted. The final response rates for men and women were 63.3% and 72.0%, respectively.

The Chi-square test or Student's t-test was used to identify significant differences between workers and non-workers in the rate or mean, respectively. The data was analyzed using the computer software package Statistical Analysis System (SAS) [14].

RESULTS

Among the subjects, 258 men held jobs and 230 men did not hold jobs at the time of the survey (Table 7.1). No significant differences were found between the two groups in terms of average age, educational background, job category in the pre-retirement period, or participation rate in various activities such as life planning, obtaining qualifications, or self-promoting activities.

Table 7.1 Comparison of basic characteristics between elderly workers and non-workers following retirement at age 60 (Men, 61 - 64 years of age)

	Workers	Non-workers	P value[1]
Number (persons)	258	230	—
Age (Mean ± SD) (years)	62.4 ± 1.1	62.6 ± 1.2	0.168
Education (above college graduate) (%)	17.1	16.6	0.516
Job category at pre-retirement (%)			
Professional, engineer (N= 37)	56.8	43.2	
Manager (N=125)	55.2	44.8	
Sales, clerical (N= 69)	44.9	55.1	0.156
Transportation, communication (N= 50)	54.0	46.0	
Manufacturing, construction (N=140)	41.4	58.6	
Security, service (N= 29)	58.6	41.4	
Activity participation (%)			
Life planning	26.0	22.7	0.400
Obtaining qualification	28.6	24.4	0.298
Self-promoting	26.7	20.9	0.133

1) Based on the chi-square test for the proportion or the t-test for the mean

Table 7.2 shows a comparison of the subjective health status between workers and non-workers. The distribution of self-rated health revealed a significantly better health status for workers than for non-workers. The percentage of the elderly who responded "poor" or "not so well" for self-rated health was lower among workers, at 6.6%, than among non-workers, at 17.0%.

The distribution of the number of symptoms also revealed a significantly better health status among workers than among non-workers. The percentage of the elderly who had two or more symptoms was significantly lower among workers, at 13.6%, than among non-workers, at 22.2%. Regarding the prevalence of individual symptoms, pain in the lower back, shoulder, or back, and blurred vision showed figures of greater than five percent in both elderly groups. Only numbness in the extremities showed a significantly lower prevalence among workers, at 3.9%, than among non-workers, at 7.9%.

The proportion of the elderly having difficulties with vision was approximately two percent in both groups. The proportion of the elderly having difficulty with hearing was significantly lower among workers, at 1.2%, than among non-workers, at 5.7%. The proportion of the elderly having difficulty in everyday life in terms of shopping, making bank deposits and withdrawals, and getting around by bus or train was low, at below 2.2%, in both groups.

Table 7.2 Comparison of subjective health status between workers and non-workers following retirement at age 60 (Men, 61 - 64 years of age)

Category of subjective health	Workers	Non-workers	P value[1]
(1) Self-rated health (%)			
Well	44.6	33.5	
Fairly well	20.5	21.7	
Average	28.3	27.8	0.002
Not so well	6.2	13.5	
Poor	0.4	3.5	
(2) Symptoms[2]			
① Number of symptoms (%)			
0	76.7	63.9	
1	9.7	13.9	0.007
2=<	13.6	22.2	
② Prevalence of symptoms (%)			
Lower back pain	10.5	14.4	0.185
Shoulder or back pain	9.3	11.8	0.647
Numbness in the extremities	3.9	7.9	0.015
Blurred vision	5.4	7.1	0.663
Fatigue	3.9	7.0	0.082
Diarrhea, constipation	2.7	4.4	0.111
Dizziness	1.9	4.0	0.307
Joint pain	5.8	3.9	0.077
Headache	2.3	3.9	0.435
Short of breath	2.7	3.5	0.099
Tinnitus	5.4	2.9	0.349
(3) Difficulties with vision or hearing (%)[3]			
Vision	1.9	2.2	0.976
Hearing	1.2	5.7	0.014
(4) Difficulties with everyday life (%)[4]			
Shopping	0.4	1.7	0.303
Conducting bank transactions	0.4	0.9	0.492
Getting around by bus or train	0.0	2.2	0.059

1) Based on the chi-square test
2) Symptoms during the past month
3) With eye glasses or contact lenses in the case of vision, and with hearing aids in the case of hearing
4) Proportion of the elderly experiencing difficulties conducting activities on their own

    The proportion of those who were seen regularly by doctors at outpatient clinics due to any diseases or disorders was 43.8% among workers, while that of non-workers was 52.2%, showing no significant difference (Table 7.3). Regarding the disease-specific outpatient rate, the only disease that showed a significant difference was kidney disease, at 5.2% for non-workers compared with 1.2% for workers. Hypertension was highest in both groups, at approximately 17%. The proportion of the elderly seen by doctors due to diabetes mellitus, back pain, and dental disease was over 5% in both groups. Approximately 6% of workers were seen by doctors due to peptic ulcer, compared with 3.5% of non-workers, showing no statistically significant difference. The proportion of the elderly who were seen by doctors due to metal disorders was approximately 1% in both groups.

Table 7.3 Comparison of percentage of the elderly being seen by doctors at outpatient clinics between workers and non-workers following retirement at age 60 (Men, 61 - 64 years of age)

	Workers	Non-workers	P value[1]
(1) Rate of elderly seen by doctors (%)	43.8	52.2	0.064
(2) Disease specific rate of the above (%)			
Hypertension	16.7	17.0	0.932
Cerebral apoplexy	0.4	1.3	0.262
Ischemic heart disease	2.3	3.9	0.311
Other cardiovascular disease	2.7	4.4	0.326
Respiratory disease	2.3	2.2	0.910
Peptic ulcer	5.8	3.5	0.224
Liver disease	2.3	3.0	0.623
Other digestive disease	0.8	1.3	0.562
Diabetes mellitus	5.4	5.7	0.913
Gout	1.9	1.3	0.582
Hyperlipidemia	4.3	2.2	0.196
Kidney disease	1.2	5.2	0.010
Skin disease	1.6	1.3	0.820
Disease in eye, ear, nose	4.3	5.2	0.620
Back pain, joint pain, neuralgia	8.9	7.4	0.540
Mental disorders	0.8	1.3	0.562
Dental disease	8.9	6.1	0.239

1) Based on the chi-square test

Table 7.4 shows a comparison of the mental health status between workers and non-workers. The mean of the CES-D scores was significantly lower for workers, at 9.9, compared with 11.1 for non-workers. The percentages of the elderly with CES-D scores of 16 or over were 5.8% and 9.1% for workers and non-workers, respectively, showing no statistically significant difference. The rates of users of drugs in terms of sleeping pills, sedatives, or depressants were 4.7% and 8.7% for workers and non-workers, respectively, showing no statistically significant difference.

Table 7.4 Comparison of mental health status between workers and non-workers following retirement at age 60 (Men, 61 - 64 years of age)

	Workers	Non-workers	P value[1]
CES-D			
Score (Mean ± SD)	9.9 ± 4.0	11.1 ± 5.0	0.011
Scores ≧ 16 (%)	5.8	9.1	0.162
Rate of users of drugs[2]	4.7	8.7	0.075

1) Based on the chi-square test in the case of proportion or the t-test in the case of mean
2) Sleeping pills, sedatives, or depressants

The rate of those who were obese or who did not receive health checkups was significantly lower among workers than among non-workers (Table 7.5). On the contrary, the rate of those with no exercise habits was significantly higher among workers than among non-workers. No significant difference was seen in the other four lifestyles, i.e., smoking, drinking, breakfast, and sleep.

Table 7.5 Comparison of health-related lifestyles between workers and non-workers following retirement at age 60 (Men, 61 - 64 years of age)

	Workers	Non-workers	P value[1]
(1) Rate of undesirable lifestyles (%)			
Smoking	38.8	46.5	0.083
Problem drinking[2]	27.1	29.6	0.551
No breakfast	2.3	4.8	0.140
No exercise[3]	72.5	63.0	0.026
Undesirable sleeping hour[4]	31.8	31.7	0.992
Obesity[5]	22.5	30.4	0.046
No health checkups	14.7	35.7	0.001
Number of undesirable lifestyles			
Distribution (%)			
0,1 (N=122)	27.5	22.2	
2 (N=177)	41.9	30.0	0.002
3 (N=119)	19.4	30.0	
4 ≦ (N= 70)	11.2	17.8	
Mean ± SD	2.1 ± 1.1	2.4 ± 1.1	0.002

1) Based on the chi-square test in the case of proportion or the t-test in the case of mean
2) Consumption of alcohol ≧ 3 units/day
3) Frequency of participation in sports or physical exercises ≧ 3/month
4) Sleeping hours <7 or ≧ 9
5) BMI (Body Mass Index) ≧ 25

The distribution of the number of undesirable lifestyles factors showed significantly better lifestyles for workers than for non-workers. The percentage of the elderly with three or more undesirable lifestyles was lower among workers, at 30.6%, than among non-workers, at 47.8%. The mean of undesirable lifestyles was significantly lower among workers, at 2.1, than among non-workers, at 2.4.

DISCUSSION

(1) Significance of this study

In Japan, the health status of workers of 61 years of age or over has not been fully clarified for several reasons. First, few studies have been conducted on the health status of elderly workers, reflecting the relatively small proportion of the elderly working after retirement. The elderly who retired from work at age 60 are considered to be covered by community health plans, for which health issues have not been examined from the standpoint of occupational health. Second, the subjects of previous studies included only elderly workers; non-workers were excluded because such studies were primarily conducted at worksites, which resulted in the studies having a low degree of internal validity due to the lack of a comparison group. Third, external validity, i.e., generalization or extrapolation, was also low, as most studies were conducted at a small number of large enterprises.

The significance of this study is due to the fact that the health status and lifestyles of male workers 61 years of age or over have been clarified in comparison with those who do not work following retirement at age 60. The two groups are considered to be comparable, as no significant difference was observed between them in terms of age, education, or job category in the pre-retirement period.

The subjects were selected by the stratified random sampling method from the corresponding age groups of the entire Japanese population. The subjects are considered to make up a representative sample of elderly workers, so the results of this study can be extrapolated to all Japanese elderly.

(2) Health status and lifestyles of elderly workers

Self-rated health is considered a comprehensive indicator of physical, mental, and social health [11]. Symptoms, vision or hearing ability, and percentage of the elderly seen by doctors at outpatient clinics are primarily considered to be indicators of physical health, while difficulty in everyday life is considered an indicator of social health. The finding that workers had a better health status than non-workers in terms of comprehensive health, physical health, mental health, and social health shows that the validity of this study is high. Therefore, the health status has been shown to be better in workers than in non-workers. It has also been clarified that a large proportion of elderly workers engage in jobs while having diseases.

Although workers were found to have a smaller number of undesirable lifestyles than non-workers, the lifestyles that were found to be better in workers were health checkups and obesity. The proportion of workers who do not engage in physical exercise regularly is extremely high compared with that of other undesirable lifestyles. For this reason, the response "no time for exercise" was the highest reported in a survey conducted for middle-aged workers [15]. The same may be true of elderly workers, although this reason was not included in the survey in this study. As the importance of physical activity for the maintenance and improvement of health has been demonstrated [17], effective workplace health-promotion programs should be implemented.

(3) Health care for elderly workers

According to a survey conducted by the Ministry of Labor [15], 65% of workers of 60 years of age or over have a disease of some kind, while in our study it was found that 44% of elderly workers are seen by physicians regularly at outpatient clinics. These findings indicate that a significant proportion of elderly workers engage in jobs while having diseases. Most of these are lifestyle-related diseases that have not been considered a target of occupational health services in the classical framework of occupational health. However, the relatively new concept of "work-related diseases" proposed by the World Health Organization in 1985 [6], includes many diseases or disorders such as lifestyle-related diseases. With work-related diseases, the work environment and the performance of work are among a number of factors that contribute significantly to the causation of a multi-factorial disease. Employers are responsible for ensuring workers' health and safety, so they may be held accountable if the health status of elderly workers with work-related diseases deteriorates due to their jobs. Health-care management for elderly workers will become a major issue in occupational health in Japan.

(4) Health promotion

To combat insufficient health care for elderly workers, more active health-care measures with the objective of promoting health will be sought [16]. The improvement of the health status of elderly workers through the improvement of lifestyles will lead not only to the prevention of work-related disease, but also to improvement in the quality of life and the containment of health-care costs. Japan has a long history of workplace health promotion. In 1979, for example, the Ministry of Labor implemented the "Silver Health Plan," a workplace health promotion scheme [7]. This program was intended to improve the physical health of middle-aged workers who had become sedentary not only in their work, but also in their daily life, by increasing their physical activity. The Silver Health Plan was revised and enlarged in 1988 to become the "Total Health Promotion Plan," in which mental health as well as physical health was targeted. These workplace health promotion programs have been shown to be effective in middle-aged workers [18,19]. Based on the experiences gained in these health promotion activities, health promotion programs that target elderly workers should be implemented.

(5) Weaknesses of this study and future tasks

This study could not determine the causal relationship among lifestyles, health status, and work, as the study employed a cross-sectional design. These relationships have not been fully clarified in the Japanese elderly, which should be a theme of a future follow-up study.

CONCLUSION

Using a representative sample of Japanese, this study clarifies for the first time that the health status and lifestyles of workers between the ages of 61 and 64 years are generally superior to those of non-workers of the same age group. It has also been clarified that a fairly large proportion of elderly workers have unhealthy lifestyles and engage in jobs while having diseases. This indicates that health promotion programs targeting elderly workers should be implemented as an occupational health measure.

ACKNOWLEDGEMENT

This study was performed through Special Coordination Funds for Promoting Science and Technology from the Ministry of Education, Culture, Sport, Science and Technology, the Japanese Government.

REFERENCES

1. Ministry of Health and Welfare, 1996, In *Annual report on health and welfare 1994-1995*, Japan International Corporation of Welfare Services.
2. Health and Welfare Statistics Association, 2001. In *Journal of health and welfare statistics*, Health and Welfare Statistics Association, (in Japanese).
3. Kumashiro, M., 1995, Productive aging with ergonomics intervention. Break down the barriers of the present hiring policy for older workers. In *The paths to productive aging*, edited by Kumashiro, M., (Taylor & Francis Ltd), pp. 1-7.
4. Prime Minister's Office, 1991, *Public opinion on life of the citizenry*, (Prime Minister's Office).
5. The Association of Employment Development for Senior Citizens, 2001. In *Handbook of labour statistics on aged society 2001*, The Association of Employment Development for Senior Citizens, (in Japanese).
6. World Health Organization, 1985, Identification and control of work-related diseases. In *WHO Technical Report Series. No. 714*.
7. Japan Industrial Safety and Health Association, 1990, *Total health promotion plan: start health promotion for mind and body*, JISHA.
8. Yamazaki, Y., 1986, *A survey on the process of retirement of middle-aged workers and their health*, Tokyo Metropolitan Institute of Labour, pp. 28-38 (in Japanese).
9. Kobayashi, R., 1990, Process of retirement of the elderly and their health problems. In *Occup Health 31(3)*, pp. 34-38 (in Japanese).
10. Fujita, D., 1998, Health consciousness of the retired people and their health

behavior, Confederation of Health Insurance Association Osaka, (in Japanese).
11. Segovia, J., Bartlett, RF., Edwards, AC., 1989, An empirical analysis of the dimensions of health status measures, In *Soc Sci Med* 29, pp. 761-768.
12. Radloff, LS., 1977, The CES-D scale. A self-reported depression for research in the general population. In *Applied Psychological Measurement 1*, pp. 385-401.
13. Belloc, NB. and Breslow, L., 1972, Relationship of physical health status and health practices, In *Prev Med 1*, pp. 409-421.
14. SAS Institute Inc, 1990, *SAS/STAT user's guide*, Version 6, fourth edition, (Cary: SAS Institute Inc).
15. Ministry of Labour, 1999, *Survey on state of employee's health 1997*, Romu Gyosei Kenkyujyo, (in Japanese).
16. Swanson, EA., Tripp-Reimer, T. and Buckwalter, K., 2001, Health promotion and disease prevention in the older adults: interventions and recommendations, (Springer Publishing Company).
17. Kampert, JB., Blair, SN., Barlow, CE. and Kohl, HW.III., 1996, Physical activity, physical fitness, and all-cause mortality: a prospective study of men and women, In *Ann Epidemiol 6*, pp. 452-457.
18. Muto, T., Hsieh, SD., Sakurai, Y., Sawada, S. and Sugisawa, H., 2000, Evaluation of workplace health promotion programs for older workers in Japan. In *Aging and work 4: Healthy and productive aging of older workers*, edited by Goedhard, WJA., (Pasmans Offsetdrukkerij b.v.), pp. 82-93.
19. Muto, T. and Yamauchi, K., 2001, Evaluation of a multicomponent workplace health promotion program conducted in Japan for improving employees' cardiovascular disease risk factors, In *Prev Med 33*, pp.571-577.

8. A Continuous Exercise Time and Psych-Physiological Reaction for a Suitable Prescriptive Exercise Program

Akiko Yamashita[1]
Mitsuyuki Kawakami and Mari Watanabe[2]
Yasumitsu Toba[3]

[1] Kanagawa University, Kanagawa, Japan
[2] Tokyo Metropolitan Institute of Technology, Tokyo, Japan
[3] Juntendo University, Tokyo, Japan

ABSTRACT

Japan has seen rapid growth in its elderly population compared with the U.S. and Europe. At the same time, birthrates have fallen, and a serious trend has emerged in Japan toward fewer children and larger populations of aged people. The number of persons aged 65 and older in particular continues to increase, and is expected to account for 25% of the total population by 2015. Therefore as one problem related to health maintenance, there is a need to consider how best to maintain and intensify bodily strength, including the strength of younger persons.

This research examines the composition and intensity of exercises applied to strengthening of individuals, as one factor in designing a prescription for appropriate exercise which aims at such maintenance and reinforcement of body strength. As the method used in this research, an experimental approach is taken, focusing on reaction in the physiological functions of the subjects resulting from different combinations of continuous exercise time and rest time. The evaluation indices used in these experiments were heart rate, expiratory volume, rating of perceived exertion, and rate of subjective fatigue. The research results indicated that, as one factor in prescription of exercise, there exists a relation between continued exercise time and rest time in suitable exercise.

Keywords: *Ergonomics, Health maintenance, Prescriptive exercise, Exercise intensity, Exercise composition*

PURPOSE OF THE RESEARCH

Average longevity is one index used to evaluate the state of health of the citizens of a nation. In Japan, the average lifetime has gradually risen, finally reaching the 80's for both men and women and making Japan the country with the longest average life spans in the world. A comparison of tendencies toward older populations among various countries indicates that, while a shift to a "graying society" occurred over periods ranging from 47 to 120 years in the U.S. and European countries, in Japan this change occurred in just 24 years. Further, it is estimated by the National Institute of Population and Social Security Research that by 2015, persons age 65 and older will account for 25% of the total population of Japan[1]. An extended lifetime is of course a blessing, but this blessing is predicated on the assumption that a healthy and active life can be lived. That is, it is desirable that the health life expectancy - the number of years over which the individual lives in a state of both physical and mental health - is also long. Aging brings with it a variety of problems, including a decline in body strength and physiological functions (sight, hearing, intelligence), as well as diseases and infirmities resulting from the aging process. If appropriate measures are taken during youth, and in middle age, a substantial effect in dealing with these problems can be anticipated, and the situation can be remedied. One such method involves exercise as a means of mitigating the decline in bodily functions; and the authors have advocated an approach which aims to maintain or increase body strength[2]. In this prescriptive exercise, it is essential that suitable exercise prescriptions based on the three principles espoused by the German biologist P.P.E. Roux (1859-1924)[3] be performed.

Therefore in this research, the continuous exercise time and rest time were examined with respect to the composition and intensity of exercise adapted to the body strength of each subject, referring to the authors' past research results, and factors for use in establishing suitable exercise prescriptions were proposed.

RESEARCH METHOD

In this research, changes in the middle-aged persons indicated in Table 8.1 when made to undergo single-load exercise on a treadmill were studied. Experiments were conducted for four different conditions: one in which exercise was continuous for a fixed duration of 20 minutes, and three other conditions in which rests were inserted intermittently into the 20 minutes of exercise time.

Table 8.1 Characteristics of Subjects (50 ages)

Subject	Age	Height (cm)	Weight (kg)	RHR	THR	METS
S1	51	154.0	51.2	77.2	123.1	7.4
S2	53	152.0	48.2	58.2	112.6	7.4
S3	51	154.0	51.6	67.2	118.1	6.2
S4	52	155.0	55.0	61.4	114.7	5.5
S5	58	152.0	57.8	62.4	112.2	7.3
S6	52	153.0	59.0	91.0	129.5	6.3
S7	58	152.0	58.0	70.4	116.2	6.6
S8	53	162.0	48.2	67.4	117.2	8.2
Average	53.5	154.3	53.6	69.4	118.0	6.9

CONTENTS OF EXPERIMENTS

The details of the experiments are presented in Table 8.2. Experiments were conducted for four different conditions of single-load exercise on a treadmill: (1) continuous running exercise for 20 minutes; (2) continuous running for 15 minutes, followed by a five-minute rest, and then another five minutes of running; (3) continuous running for 10 minutes, followed by a five-minute rest, then five minutes of running, and another five minutes of rest before five more minutes of running; and, (4) alternation of five minutes of running, and five minutes of rest. In order to observe changes in heart rate and expiratory volume as a result of the exercise load under all four conditions, five minutes of quiet rest before, and 10 minutes of quiet rest after the exercise were imposed. To measure the time of quiet rest before, rest time during exercise, and quiet rest after the exercise, the treadmill was stopped, and the subject was made to assume a sitting posture in a chair and rest quietly on the treadmill during measurements.

Table 8.2 Contents of experiment

Experimental conditions	Condition 1	Pre-exercise rest (5 min)	Load (Exercise) (20 min) - Post-exercise rest (10 min)
	Condition 2	Pre-exercise rest (5 min)	Load (Exercise) (15 min) - rest (5 min) - Load (Exercise) (5 min) - Post-exercise rest (10 min)
	Condition 3	Pre-exercise rest (5 min)	Load (Exercise) (10 min) - rest (5 min) - Load (Exercise) (5 min) - rest (5 min) - Load (Exercise) (5 min) - Post-exercise rest (10 min)
	Condition 4	Pre-exercise rest (5 min)	Load (Exercise) (5 min) - rest (5 min) - Load (Exercise) (5 min) - rest (5 min) - Load (Exercise) (5 min) - rest (5 min) - Load (Exercise) (5 min) - Post-exercise rest (10 min)

Evaluation indices used in experiments as physiological indices were, as shown in Table 8.3, (1) heart rate (HR) and (2) expiratory volume (VO_2, VCO_2, VE), and as subjective indices, (1) Borg's subjective exercise intensity (RPE: rating of perceived exertion), and (2) the subjective fatigue. The heart rate and expiratory volume were measured continuously from the start of the initial quiet rest period until the end of the final quiet rest period. The RPE was recorded immediately after the end of exercise (immediately before the final quiet rest period). The subjective fatigue was recorded upon starting the initial quiet rest period before the experiment, and after ending the final quiet rest period immediately after conclusion of the experiment. The settings for exercise time were adopted based on the definition of cardiac fatigue by Ishii, Toyota and Yamaguchi, 1983[4], and from the definition of training of K. & J. Kagaya, 1983[5].

Table 8.3 Experimental Conditions

Evaluation indices	Physiological indices: Heart rate (HR), expiratory volume (VO_2, VCO_2, VE) Subjective indices: RPE (rating of perceived exertion), proportion of subjective fatigue symptoms
Subjects	Eight healthy females in their 50's
Exercise time	Condition 1: 35 min Condition 2: 40 min Condition 3: 45 min Condition 4: 50 min (Including Pre-exercise rest time, Post-exercise rest time and Rest time between exercise)
Experiment environment	Average room temperature: 26°C Average humidity: 55%

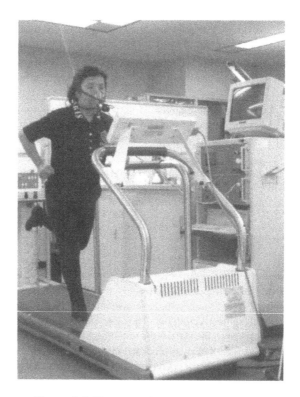

Figure 8.1 Example of Experimental Situation

Exercise intensity was calculated from the THR (middle-aged persons: intensity level 50%) computed using eq. (1) from the target heart rate (THR) advocated by Karvonen, 1957[6], and from eq. (2) of Kawakami, Inagaki, Yamashita & Ukai, 1999[7] predicting the heart rate based on the bodily characteristics of each patient, using eq. (4) and (6), and from an equation for the energy metabolism rate (RMR: relative metabolism rate).

(1) $THR = (1 - 0.6) \times HRrest + 0.6 \times (220 - Age)$

(2) $HRsubmax = \alpha \times RMR + HRrest$

(3) $= \dfrac{Sb \times (220\ Age\ HRrest)}{4.15 \times 10^{-3} \times Weight \times \%VC + 2.50}$

(4) $Sb = 74.49 \times (Height)^{0.725} \times (Weight)^{0.425}$

(5) $RMR = 1.20 \times (METS - 1)$

Here eq. (1) is used, with HR_{submax} in eq. (2) set so as to be equal to THR, and eq. (2) is modified to obtain eq. (6):

(6) $THR = \alpha \times RMR + HRrest$

EXPERIMENTAL RESULTS

1. Physiological Indices

(1) Heart Rate

Changes in heart rate classified by conditions are shown in Fig. 8.2. The changes in heart rate shown in Fig. 8.2 give the change in heart rate relative to the average heart rate during the post-exercise rest of experimental condition 1. Fig. 8.2 indicates that recovery of heart rate is fastest for condition 3, and recovers in the order of condition 4, condition 2. It is thought that compared with the continuous running of condition 1, condition 3 (exercise for 10 minutes, rest for five minutes, exercise for five minutes, rest 5, exercise 5) is exercise under reasonable conditions. This data was used in a ternary variance analysis, taking the experiment conditions, subject, and elapsed time as factors. As a result, significant differences were observed, with 99% reliability, between the conditions. In this analysis, significant differences were observed between the subject factors, and so this was considered in examining differences in average values associated with the data. The results appear in Table 8.4 from the table, significant differences were observed between condition 1 and condition 3, condition 2 and condition 3, condition 2 and condition 4, and condition 3 and condition 4 with 99% reliability, and between condition 1 and condition 4 with 95% reliability. From these results and the rate of change in the heart rate after the post-exercise rest time for middle-aged persons in Fig. 8.2, condition 3 is thought to be the most desirable condition.

Table 8.4 Result of t-test

Factors	Freedom	t	Approval
Conditions 1, 2	59	0.01	Accepted
Conditions 1, 3	59	9.32**	Rejected
Conditions 1, 4	59	2.55*	Rejected
Conditions 2, 3	59	4.88**	Rejected
Conditions 2, 4	59	2.76**	Rejected
Conditions 3, 4	59	2.94**	Rejected

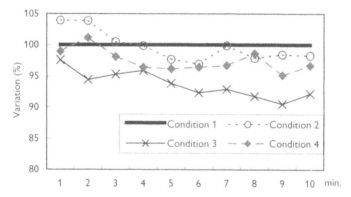

Figure 8.2 A variation of HR on each experimental conditions (Post rest time)

(2) Expiratory Volume (VO2, VCO2, VE)

In an approach similar to that of the above analysis, the average during post-exercise rest time of VO2, VCO2, VE for experiment condition 1 was set to 100, and the VO2, VCO2 and VE during the post-exercise rest time for each of the second, third, and fourth conditions were evaluated. The results are shown in Figs. 8.3, 8.4 and 8.5. Compared with condition 1, there is a declining tendency in all cases for 3 conditions, in which a rest is inserted. As with the heart rate analysis, these results were used in a ternary variance analysis, taking the experiment condition, subject, and elapsed time as factors. The results revealed significant differences in each of VO2, VCO2, and VE with 99% reliability. In these results, significant differences were observed among subject factors, and so using an approach similar to that described above, differences in average values were examined. The results appear in Table 8.5, Table 8.6 and Table 8.7.

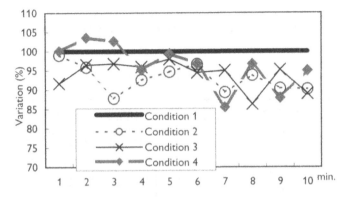

Figure 8.3 A variation of VO_2 on experimental

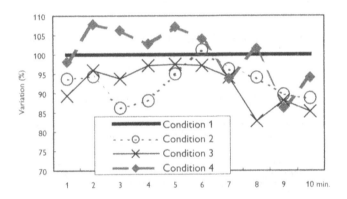

Figure 8.4 A variation of VCO_2 on experimental

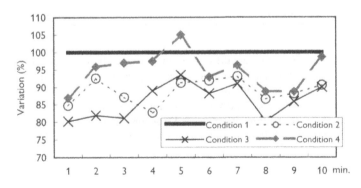

Figure 8.5 A variation of VE on experimental each conditions (Post rest time)

From Table 8.5, significant differences were observed in VO2 between condition 1 and condition 2 with 99% reliability, and between conditions 1 and 3 with 95% reliability.

Table 8.5 Result of t-test (VO_2)

Factors	Freedom	t value	Approval
Conditions 1, 2	39	4.56**	Rejected
Conditions 1, 3	39	2.70*	Rejected
Conditions 1, 4	39	1.37	Accepted
Conditions 2, 3	39	0.23	Accepted
Conditions 2, 4	39	1.83	Accepted
Conditions 3, 4	39	1.38	Accepted

From Table 8.6, significant differences in VCO2 were observed between conditions 1 and 2, conditions 1 and 3, conditions 2 and 4, and conditions 3 and 4 with 99% reliability.

Table 8.6 Result of t-test (VCO_2)

Factors	Freedom	t value	Approval
Conditions 1, 2	39	3.73**	Rejected
Conditions 1, 3	39	3.42**	Rejected
Conditions 1, 4	39	0.27	Accepted
Conditions 2, 3	39	0.41	Accepted
Conditions 2, 4	39	3.69**	Rejected
Conditions 3, 4	39	3.74**	Rejected

From Table 8.7, significant differences in VE were observed between conditions 1 and 2, conditions 1 and 3, conditions 2 and 4, and conditions 3 and 4 with 99% reliability, and between conditions 1 and 4, and conditions 2 and 3 with 95% reliability. Based on these results and the results for the rates of change of VO2, VCO2 and VE in the post-exercise rest time, it can be said that condition 2 is preferable with respect to VO2, and that condition 3 is preferable with respect to VCO2 and VE.

Table 8.7 Result of t-test (VE)

Factors	Freedom	t value	Approval
Conditions 1, 2	39	4.46**	Rejected
Conditions 1, 3	39	5.53**	Rejected
Conditions 1, 4	39	2.37*	Rejected
Conditions 2, 3	39	2.36*	Rejected
Conditions 2, 4	39	2.89**	Rejected
Conditions 3, 4	39	5.33**	Rejected

Table 8.8 Result of Experiment

Evaluation indices ╲ Subjects		Middle-aged and elderly person
Heart rate (HR)		Condition 3
expiratory volume	VO_2	Condition 2
	VCO_2	Condition 3
	VE	Condition 3
RPE (rating of perceived exertion)		Condition 4
proportion of subjective fatigue symptoms		Condition 4

2. Subjective Evaluation Indices

(1) RPE: Rating of Perceived Exertion

The RPE, or rating of perceived exertion, is the intensity of exertion felt by the subject in exercise; the higher the number indicated by the subject, the greater the sense of fatigue induced by the exercise. Fig. 8.6 compares figures for the rating of perceived exertion for each of the conditions. Based on Fig. 8.6, it can be said that condition 4, in which the greatest number of rest periods are inserted between exercise periods, is the most preferable of the four conditions, and results in the lowest sense of fatigue.

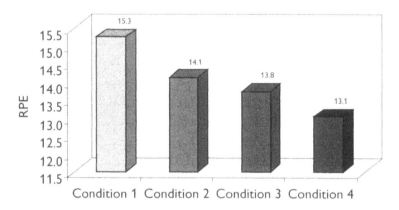

Figure 8.6 Value of RPE on each conditions

(2) Proportion Surveyed Subjective Symptom

Fig. 8.7 shows changes in the rate of complaints of perceived fatigue, arranged by groups. Observing changes in overall rates of complaints as seen in the results of Fig. 8.7, high rates of complaints of fatigue are seen for conditions 1 and 2. On the other hand, complaint rates are seen to be lower for conditions 3 and 4. Comparison of conditions 3 and 4 indicates that complaint rates are lower for condition. 4, with the greater frequency of rests between exercise. From this it is concluded that condition 4 is the most preferable.

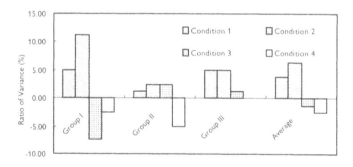

Figure 8.7 A Ratio of Variance of Survey of Subjective Symptom

CONCLUSIONS

In this research, continuous exercise time and psycho-physiological reactions to suitable prescriptive exercise programs were studied to identify factors for establishing suitable prescriptive exercise programs for middle-aged persons.

As a result of this research, the following conclusions were obtained.

(1) As one factor in exercise prescriptions for the middle-aged subjects, there exists a relation between continuous exercise time and rest time in suitable exercise.
(2) When using average exercise intensity (METS: 6.9) as an overall evaluation index, an exercise prescription in which 5 minutes of rest are inserted after every 5 to 10 minutes of exercise is preferable.

9. The Effect of Aging on Lipid Metabolism and Aerobic Ability

Yasumitsu Toba[1], Akiko Yamashita[2], Mitsuyuki Kawakami[3], Shigenobu Arai[4], Shizuo Sakamoto[1], and Toshihiko Iijima[1]

[1] Department of Life and Medicine, Juntendo University School of Medicine Urayasu Hospital, Urayasu, Chiba, JAPAN
[2] Department of Sports Science and Physical Education, Kanagawa University, Yokohama, Kanagawa, JAPAN
[3] Tokyo Metropolitan Institute of Technology, Tokyo, JAPAN
[4] Department of Exercise Science and Human Performance, National Defense Academy of Japan, Yokosuka, Kanawaga, JAPAN

ABSTRACT

The purpose of this study was to find out the effects of aging on lipid metabolism by using Maximal Lipid Combustion Rate (MLCR), and on aerobic ability during exercise for three different age -groups. All subjects in this study were Japanese male workers (n = 206, range of age: 20 - 74 years old) who received health check-ups prior to performing the exercise stress test. They were categorized in three different groups according to age: (1) Age 20 - 39 (range of 20 - 39 years old: n = 53), (2) Age 40 - 59 (range of 40 - 59 years old: n = 100), (3) Age 60 - 74 (range of 60 - 74 years old: n = 53). Electrically braked cycle ergometer and gas analyzing system were used for this exercise stress test with gas analyzing (Ramp method: 20watt/min). Gas samples were collected by the Breath-by-Breath method, and analysis was done of Peak Vo2 (Peak Vo2), Peak Heart Rate (Peak HR), Maximal Lipid Combustion Rate (MLCR), Vo2 at Maximal Lipid Combustion Rate (MLCR-Vo2), and Vo2 at Anaerobic threshold point (AT-Vo2). Percentage rate of AT-Vo2 and Peak Vo2 (AT-Vo2 / Peak Vo2), Percentage rate of MLCR-Vo2 and Peak Vo2 (MLCR-Vo2 / Peak Vo2) were also calculated, and data values of each group were compared and analyzed to find the differences between age groups. Endurance ability is shown to lower with aging from the standpoint of data of Peak HR, Peak Vo2, AT-Vo2, MLCR-Vo2. However, the percentage rates of AT-Vo2 / Peak Vo2 and MLCR-Vo2 / Peak Vo2 increased with aging. It is suggested that any kind of function similar to homeostasis in the human body, that maintains Aerobic ability or helps to prevent it from decreasing, and lipid metabolism activation might be useful for the elderly.

Keywords: *Aging, Lipid, Aerobic ability, Exercise, Male worker*

INTRODUCTION

The frequency of abnormal lipid metabolism, including obesity and other life-related diseases, is high even among the many modern diseases from which humankind suffers, and in particular, Hypercholesterolemia is gradually increasing according to a report by Isomura [6]. Hypercholesterolemia, Hyperlipemia, etc. are thought of as three major coronary risk factors, and it is important to measure them. It is thought that exercise therapy is an effective method, along with diet treatment [5,10], for controlling not only abnormal lipid metabolism such as Hyperlipemia and Hypercholesterolemia, but also for other Life-related diseases. In particular, exercise therapy is important as the basis of diabetes treatment [7]. If more lipids are used as the energy source during exercise, it is expected that an effective improvement could be made in lipid metabolism. It is reported that the functional capacity of humans declines 0.75% to 1.0% per year after the age of 30 [13], and other studies also report decreasing ability related to age [15]. Actually, it is well known that the physical ability of the human body declines with aging. However, it is necessary to live a healthy life in order to enjoy good health in old age, which requires a good balance of "Nutrition, Rest, and Exercise". In particular, it is thought that people today need more exercise to keep healthy and avoid life-related diseases. Therefore, it is important to consider the relationship between aging and the human body.

PURPOSE

The purpose of this study was to find out the effects of aging on lipid metabolism by using Maximal Lipid Combustion Rate (MLCR) and Aerobic ability during exercise.

SUBJECTS

All subjects in this study were Japanese male workers (n = 206, range of Age: 20 - 74 years old) who were treated at a university hospital. They were provided with documentation in order to make written informed consent, their health condition was checked, and physical characteristics were measured prior to performing the exercise stress test at Juntendo University School of Medicine Urayasu Hospital in Urayasu, Japan. Some subjects were taking medication for Obesity, Hyperlipemia, Hyperuricemia, Fatty Liver, and Diabetes. The subjects did not have the habit of regular physical or recreational activity.

Subjects were categorized in the following three different groups by age level. Table 9.1 shows physical characteristic of subjects in each age group.

1) Age 20 - 39 (range of 20 - 39 years old): n = 53, Average age: 34.1 ± 5.3 years old, Average height: 171.5 ± 5.9 cm, Average body weight: 79.1 ± 16.5 kg, Average rate of body fat: 26.0 ± 8.1 %, Average body mass index (BMI): 26.9 ± 5.8
2) Age 40 - 59 (range of 40 - 59 years old): n = 100, Average age: 50.5 ± 5.7 years old, Average height: 168.9 ± 5.8 cm, Average body weight: 70.3 ± 10.7 kg, Average rate of body fat: 22.5 ± 5.4%, Average body mass index (BMI): 24.6 ± 3.1

3) Age 60 - 74 (range of 60 - 74 years old): n = 53, Average age: 64.1 ± 3.6 years old, Average height: 164.8 ± 5.8 cm, Average body weight: 65.9 ± 10.0 kg, Average rate of body fat: 20.7 ± 5.2 %, Average body mass index (BMI): 24.2 ± 3.0

Table 9.1 Physical characteristic of subjects (n = 206)

Age group	Age (yrs.)	Height (cm)	Weight (kg)	Body Fat (%)	BMI
20ies - 30ies (n = 53)	31.4 ± 5.3	171.5 ± 5.9	79.1 ± 16.5	26.0 ± 8.1	26.9 ± 5.8
40ies - 50ies (n = 100)	50.5 ± 5.7	168.9 ± 5.8	70.3 ± 10.7	22.5 ± 5.4	24.6 ± 3.1
60ies - 70ies (n = 53)	64.1 ± 3.6	164.8 ± 5.8	65.9 ± 10.0	20.7 ± 5.2	24.2 ± 3.0

METHODS AND MATERIALS

All subjects in each group were provided with written information on this study, had their health checked prior to the test, and then performed an exercise stress test (by Ramp method: 20 watt/min.) by electrically braked cycle ergometer (ERGOMETER 232XL, Combi Co., Tokyo, Japan). Gas samples were collected by "Breath-by-Breath" method with subjects at rest, seated on ergometer, and throughout the exercise period. Data of Peak Vo2 (Peak Vo2), Peak Heart Rate (Peak HR), Maximal Lipid Combustion Rate (MLCR), Vo2 at Maximal Lipid Combustion Rate point (MLCR-Vo2), and Vo2 at Anaerobic threshold point (AT-Vo2) were analyzed by a Gas analyzing system (Aeromonitor AE-280, Minato Medical Science Co., Osaka, Japan [14]. Percentage rate of AT-Vo2 in Peak Vo2(AT-Vo2 / Peak Vo2), Percentage rate of MLCR-Vo2 in Peak Vo2 (MLCR-Vo2 / Peak Vo2) were also calculated, and data values for each group were compared and analyzed to study the differences between age groups.

STATISTICAL ANALYSES

Paired Student's t-tests were used to test for significance between groups. Values are expressed as mean ± SD, and the 0.05 level of confidence was accepted for statistical significance.

RESULTS

Peak Vo2 (Peak Vo2)

Table 9.2-1 and Fig. 9.1 provide Peak Vo2 value of each age group. Peak Vo2 value of 20s - 30s age group: 27.87 ± 8.80 ml/kg/min was the highest, 25.15 ± 5.55 ml/kg/min of the 40s - 50s age group was next after the 20s - 30s, and the lowest Peak Vo2 value was 23.02 ± 3.78 ml/kg/min of the 60s - 70s age group. There were significant differences between the 20s - 30s age group and 40s - 50s (P < 0.05) and 60s -70s (P

< 0.01) age groups. There were also significant differences between the 40s - 50s age group and the 60s - 70s (P < 0.05) age group.

Table 9.2-1 Summary of data

(n = 206)

Age group	Peak Vo2 (ml/kg/min)	Peak HR (bpm)	MLCR (mg/min)	MLCR-Vo2 (ml/kg/min)
20ies-30ies	27.87 ± 8.80	163.6 ± 16.4	320.1 ± 132.9	13.10 ± 3.90
40ies-50ies	25.15 ± 5.55	151.7 ± 13.8	250.1 ± 102.4	12.42 ± 3.09
60ies-70ies	23.02 ± 3.78	142.9 ± 14.1	235.4 ± 123.5	12.31 ± 2.57

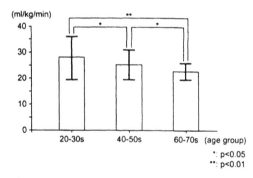

Figure 9.1 Comparison of Peak Vo$_2$ between Age groups: (Peak Vo2)

Peak Heart Rate (Peak HR)

Peak Heart Rate values for each group are shown in Table 9.2-1 and Fig. 9.2. Peak Heart Rate value of 20s - 30s age group was 163.6 ± 16.4 beat/min, and was the highest value. 151.7 ± 13.8 beat/min of the 40s - 50s age group was second, and the lowest value was 142.9 ± 14.1 beat/min of the 60s - 70s age group. There were no significant differences between groups.

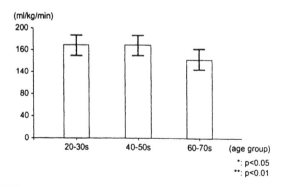

Figure 9.2 Comparison of Peak Heart Rate between Age groups: (Peak HR)

Maximal Lipid Combustion Rate (MLCR)

Table 9.2-1 and Fig. 9.3 show Maximal Lipid Combustion Rate (MLCR) for each age group. MLCR value was high in the order of 235.4 ± 123.5 mg/min (60s - 70s age group), 250.1 ± 102.4 mg/min (40s - 50s age group), 320.1 ± 132.9 mg/min (20s - 30s age group). There were statistically significant differences between the 20s - 30s age group and 40s - 50s age group (P < 0.01) and 60s - 70s age group (P < 0.01). The trend for the MLCR value was a decrease with aging.

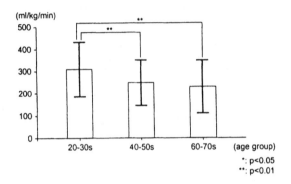

Figure 9.3   Comparison of Maximal Lipid Combustion Rate between Age groups: (MLCR)

Vo2 at Maximal Lipid Combustion Rate Point (MLCR-Vo2)

The order of volume of MLCR-Vo2 was 60s - 70s age group (12.31 ± 2.57 ml/kg/min) < 40s - 50s age group (12.42 ± 3.09 ml/kg/min) < 20s - 30s age group (13.10 ± 3.90 ml/kg/min). Even this MLCR-Vo2 value tended to decrease with aging. However, there were no statistically significant differences between the groups. Data for MLCR-Vo2 are provided in Fig. 9.4 and Table 9.2-1.

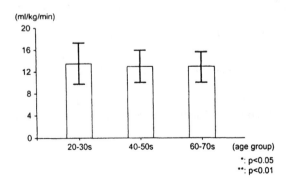

Figure 9.4   Comparison of VO_2 at Maximal Lipid Combustion Rate Point between Age groups: (MLCR-Vo2)

Vo2 at Anaerobic Threshold Point (AT-Vo2)

Table 9.2-2 and Fig. 9.5 provide the AT-Vo2 value for each group. 13.71 ± 4.89 ml/kg/min (20s - 30s age group) was the highest value, 12.83 ± 3.80 ml/kg/min. (40s - 50s age group) was second, and the lowest value was 12.61 ± 2.73 ml/kg/min. (60s - 70s age group).

There were no significant differences between groups. However, this AT-Vo2 value also tended to decrease with aging.

Table 9.2-2 Summary of data

(n = 206)

Age group	AT-Vo2 (ml/kg/min)	AT-Vo2/ Peak Vo2 (%)	MLCR-Vo2/ Peak Vo2 (%)
20ies-30ies	13.71 ± 4.89	46.9 ± 9.1	47.8 ± 8.3
40ies-50ies	12.83 ± 3.80	51.1 ± 8.9	49.7 ± 7.5
60ies-70ies	12.61 ± 2.73	54.9 ± 7.4	54.1 ± 10.6

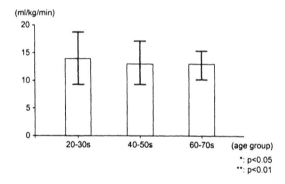

Figure 9.5 Comparison of Vo2 at AT point between Age groups: (AT-Vo2)

Percentage of AT-Vo2 in Peak Vo2 (AT-Vo2 / Peak Vo2)

Percentage values of AT-Vo2 / Peak Vo2 are shown in Table 9.2-2 and Fig. 9.6. There were significant differences of P < 0.01 between the 20s - 30s age group (46.9 ± 9.1 %) and 60s - 70s age group (54.9 ± 7.4 %), and also differences of P < 0.01 between the 40s - 50s age group (51.1 ± 8.9 %) and 60s - 70s age group (54.9 ± 7.4 %). It was an interesting phenomenon that the value of this percentage rate increased with aging.

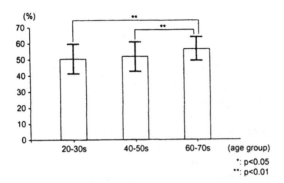

Figure 9.6   Comparison of Percentage of Vo2 at AT point (AT-Vo2) in Peak Vo2
(Peak Vo2) between Age groups: (AT-Vo2/Peak Vo2)

Percentage of MLCR-Vo2 in Peak Vo2 (MLCR-Vo2 / Peak Vo2)

Table 9.2-2 and Fig. 9.7 show that the value of percentage of 60s - 70s age group
(54.1 ± 10.6 %) is significantly greater ($P < 0.01$) than the values of the 20s - 30s
(47.8 ± 8.3 %) and 40s-50s (49.7 ± 7.5 %) age groups. The value for the 40s - 50s age
group is greater than that of the 20s - 30s age group, but there were no significant
differences in the statistical analysis.

An interesting phenomenon appeared, namely, the value of percentage rate
increased with the aging.

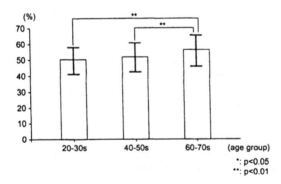

Figure 9.7   Comparison of Percentage of Vo2 at Maximal Lipid Combustion
Rate point at (MLCR-Vo2) in Peak Vo2 (Peak Vo2) between Age
groups: (MLCR-Vo2/Peak Vo2)

DISCUSSION

Results of Peak Vo2 value in this study proved that the value decreases with aging (Fig. 9.1 and Table 9.2-1). It is well known and normal that physical function and ability of the human body decrease with aging [15]. Many studies report that muscle strength, other physical functions, and Vo2max, which is one of the most important indications of endurance ability, decline or decrease with aging [4,5,9,12], Buskirk, E.R., et al. reported that Vo2max in males decreased 0.40 - 0.50 ml/kg/min per year [2], and other studies also noted that Functional capacity of humans declines 0.75% to 1.0% per year after the age of 30 [13]. The relationship of Peak Vo2 and Vo2max is very close. Peak Vo2 and Peak Heart Rate are also indicators of physiological response of the human body. Therefore, it is natural, according to previous research, that decreasing Peak Vo2 and Peak Heart Rate relate to aging.

Maximal Lipid Combustion Rate (MLCR) is used as an indication of the movement of lipid metabolism.(Fig. 9.3 and Table 9.2-1). Advanced research uses Respiratory Quotient (RQ) and examines the occupying ratio of lipid and carbohydrate in energy for exercise [1]. The measurement of lipid burning (combustion) rate in this study was applied, and the method of numerical formula was input to a computer analyzing system [14] in order to examine and output a lipid burning curve. The data of Maximal Lipid Combustion Rate (MLCR) showed the differences in lipid burning during exercise for each age group, and a lower MLCR value was shown to be related to aging. The reason is the influence of declining body function, and this factor would be based on activating of metabolism. Any function improves if the metabolism is activated. However, metabolism and other functional ability would be adversely affected by aging. Vo2 value at MLCR point and at AT point also would decrease with aging. It is normal that the absolute values of Vo2 at MLCR point and AT point were influenced by the decline in functional capacity due to aging. Actually, it is known that one of main reasons for the decline of muscle strength is caused by decrease in muscle volume related to age [8]; walking ability also related to muscle strength in the lower part of the body. Values of Vo2 at MLCR point and Vo2 at AT point that were on the load level of walking would be related to metabolism and physical function for each age.

The percentage rate of Vo2 value at AT point in Peak Vo2 and at MLCR point in peak Vo2 (Fig. 9.6 and 9.7). Generally, it is known that human function and other systems decline with aging [2,3,8,12,13,15]. However, both of the above factors indicated that their value increased with aging. AT point means that it is the switch or transfer point from anaerobic energy supply to Aerobic energy system [11]; in other words, this is the max point of load to perform at Aerobic working level. The date of Vo2 at AT point against Peak Vo2 in Fig. 9.6 and Table 9.2-2 would mean that some kind of function might be controlling to maintain aerobic ability as the human body ages. Absolute value of MLCR (Fig. 9.3) decreased with aging. However, the phenomenon of percentage of Vo2 at MLCR point against Peak Vo2 is different from the case of MLCR.

It suggested that the human body would keep the higher activation of lipid metabolism to control fat when the body ages. Finally, both results suggested that any kind of homeostasis-like function in the human body that helps to maintain Aerobic ability or prevent decreasing of it, and keep lipid activation, would be useful.

CONCLUSION

The purpose of this study was to find the differences in the movement of lipid metabolism by using Maximal Lipid Combustion Rate (MLCR) and Aerobic ability during exercise related to aging. The endurance ability of each group declined with aging according to these data. However, it is suggested that any kind of homeostasis-like function in the human body that maintains or helps to prevent decreasing Aerobic ability, and lipid metabolism activation, might be useful even in older people.

REFERENCES

1. Bursztein, S. *et al.*, 1989, Energy metabolism, indirect calorimetry, and nutrition, Williams & Wilkins, pp .27-57
2. Buskirk, E.R. *et al.*, 1987, Age and aerobic power. In *The rate of change in men and woman, Federation Proc, 46*, pp. 1824-1829.
3. Dehn, MM., 1972, Longitudinal variations in maximal oxygen intake with age and activity. In *J. Appl. Physiol, 33*, pp. 805-807.
4. Frontera, MM., 1991, Cross-sectional study of muscle strength and mass in 45-to 78 yr-old men and women. In *J. Appl. Physiol, 71*, pp. 644-650.
5. Holloszy, JO., 1964, Effects of six month program of endurance exercise on the serum lipid of muddle-aged men, Am. In *J. Cardiol. 14*, p. 253.
6. Isomura, K., 1989, Eiyousesshu no kin-nen no hensen to Jyunkanki-shikkan (Change of Nutritional intake in recent years and Circulatory disease), (Suzuken Centre Publishers), pp. 55-72.
7. Kemmer, FW., *et al.*, 1993, Exercise and diabetes mellitus: physical activity as a part of daily life and its role in the treatment of diabetic patients. In *Int. J Sports Med 4*, pp. 77-88.
8. Kim, J.D., *et al.*, 2000, Relationship between reduction of hip joint and thigh muscle and walking ability in elderly people, The Japanese J. of Phys. In *Fit and Sports. Med. 49(5)*, pp. 589-596.
9. Miyamoto, H., *et al.*, 2000, The effect of habitual jogging on the decrease of maximal oxygen consumption with aging. In *J of Japanese Society of Clinical Sports Med, 9-2*, pp. 240-245.
10. Nakamura, H., 1989, Hypelipemia, Symposium: Sports Medicine. In *The Saishin-igaku 43(10)*, pp. 2248-2252.
11. Nakano, S., 2001, Ilustration; The merit and demerits of physical exercise on human body, pp. 149-151, (The Ishiyaku-shuppan publishers, Inc).
12. Porter, M.M., 1995, J. Aging of human muscle: structure, function and adaptability. Scand. J. Med. In *Sci.Sports 5*, pp. 129-142.
13. Schilke, JM., 1991, Slowing the aging process with physical activity. In *J Gerontol Nurs*, Jun, 17(6), pp. 4-8.
14. Shimoda, *et al.*, 1984, Evaluation of a System for On-line analysis of Vo2 and Vco2 for Clinical Application. In *Anesthesiology, 61*, pp. 311-314.
15. Shiozaki, T. *et al.*, 1998, Characteristics of muscle oxygenation in elderly men determined by near infrared spectroscopy. In *The Japanese J. of Phys. Fit and Sports. Med.47(4)*, pp. 393-400.

10. For Whom is a Disability Pension a Good Solution When Musculoskeletal Disorders Prevent Work?

Lena Edén[1], Göran Ejlertsson[1], Jan Petersson[2]

[1] Department of Health Sciences, Kristianstad University, Sweden
[2] Department of Social Work, Lund University, Sweden

ABSTRACT

When work ability is permanently reduced, individuals in industrialised countries have the chance to receive a disability pension. It is meant to guarantee an economic basis for those unable to earn their living. A disability pension might entail positive as well as negative consequences for the individual.

The aim of this study was to find determinants of improving quality of life (QL) among disability pensioners with musculoskeletal disorders (DPs) and to estimate the importance of subjective health for QL. Questionnaires were distributed 1-6 years after retirement and two years later to 352 DPs. In addition to unchanged or improved subjective health and ADL status, improvement in QL since retirement was related to female gender, age nearing statutory old-age pension, unemployment preceding the retirement, the view that a disability pension is all right today, and satisfaction with social network and leisure activities. Among these explanatory variables, gender and health status explained improvement in QL at the follow-up two years later. Changes in living conditions imposed by retirement, and not life as a retiree, were seen as crucial for improvement in QL among DPs.

Keywords: *Disability pension, Early retirement, Musculoskeletal disorders, Quality of life*

INTRODUCTION

In Sweden, as in other industrialised countries, many workers - especially older ones - leave the labour market permanently in advance of the statutory pension age. If work ability is reduced due to chronic ill health or injury, a disability pension is considered and may be granted as a full-time or part-time benefit depending on the degree of the incapacity. The rate of early retired pensioners due to disability (DPs) increased in the population aged 16-64 years in Sweden from 3.7% in 1970 to 7.6% in 1999. Increases in the rate of DPs have been seen in other industrialised countries as well [1-3] and different models to explain this increase have been developed [4-5]. According to the pull-model, the social security system, with increasing generosity regarding financial compensations and opportunities to retire early, is attractive for the individual. The push-model focuses on the conditions of the labour market, where increasing demands eliminate workers who fail to achieve the set standards. Pull factors could be seen as emanating from positive, and push factors from negative, considerations [6]. Individual factors such as age, educational level and health status are of importance when a disability pension is considered. Some emphasise the "political economy" perspective [4] and add a systems model, whereby variations in the rate of early retirees are explained by changes in the rules of the welfare system or practices among civil servants at the local social insurance offices [7].

The paradox in the drop of the age of exit from gainful work parallel to an increase in health resources for aged people in Western Europe has been stressed [3, 8]. The general explanation is that access criteria have been somewhat relaxed, i.e., the right to a pension was strengthened [9].

The vast majority of disability pensioners (72%) in the late 1970s did not want to go back to work, mainly due to health problems which made them consider work as a bad alternative [10]. A disability pension was thus increasingly seen as a social right by some individuals, while others most likely still felt that they were expelled from the labour market against their will. This combination of entitlement and compulsion might affect the quality of life (QL) among DPs in the same way as has been shown in a study of voluntary and involuntary early exits from the labour market [6]. In a previous article [11], we have shown that individuals who were forced to take a disability pension against their own wishes ran the risk of having poor QL in the future. The same was true for individuals who had been working or out of work less than one year during the five years preceding the retirement decision - a result that indicates that a disability pension is seen as a relief for the individual who recently experienced a long period of unemployment.

Other indicators of poor QL among DPs we found [11] were: being an immigrant, having a poor self-rated health status, experiencing deteriorating ADL capacity, suffering from a high level of neurotic symptoms, lacking opportunities to indulge in some leisure time activity, having a poor self-image and being discontented with one's financial situation and social network.

Getting an objective disease diagnosis by a physician is seldom related to poor QL [12], whereas the subjective health status is of great importance for QL in the general population [12-15] as well as among DPs in our study [11, 16]. The health status of an individual might be influenced positively [17] or negatively [18] by a disability pension. Our study reveals that women ≥55 years of age declare that their subjective health status improved after a disability pension, while younger women and men of all ages run the risk of a deterioration, and immigrants were found to have an exposed position concerning change in health status as DPs [19].

A disability pension is part of the welfare system aiming at economic security for the individual. However, a welfare approach ought to include more than economic/materialistic considerations; welfare from the individual's point of view should be considered as the possibility to be satisfied with different aspects of life, i.e., to experience good QL.

The aim of the present paper was to identify determinants of a perceived improvement in quality of life among disability pensioners suffering from musculoskeletal disorders. A further aim was to estimate the importance of subjective health for quality of life among the disability pensioners.

MATERIAL AND METHODS

The population-based study was performed in Kristianstad municipality in southern Sweden, which has about 70,000 inhabitants. The study group consisted of all 450 persons aged 25-59 years who were granted a full-time early retirement pension during the period 1986-1990 due to disorders of the musculoskeletal system. They were defined by all diagnoses in chapter 13 (diseases of the musculoskeletal system and connective tissues), and diagnoses within the musculoskeletal system in chapter 17 (injuries), in the ninth revision of the international classification of diseases (WHO: ICD-9). They constituted about 50% of all newly granted disability pensions in the studied area during the period.

In January 1992, a questionnaire was sent to all DPs in the study group. The response rate was 83.6%. Two years later a second questionnaire was sent to all DPs who answered the first one. The response rate this time was 95.1%. The age and sex distribution of the participants in the 1994 survey is described in Table 10.1. The study group and the non-response analysis have been described in greater detail previously [20].

Table 10.1 Age and sex distribution among the participants in the 1994 survey (%)

	Males (n=111)	Females (n=241)
- 44 years	4.5	6.6
45 - 54 years	17.1	26.6
55 - years	78.4	66.8

The DPs were aged 25-59 years at the time of retirement.
The questionnaire was sent out 3-8 years after retirement.

Quality of life, definition

Quality of life here is referred to as the individual's evaluation of his/her life content, i.e., global QL, and is based on self-report. This is in accordance with how several other authors define QL [e.g. 14-15, 21]. The concept of global QL here is dealt with from two aspects: present quality of life (PQL), which reflects the current life situation, and change in quality of life (CQL) since retirement (the 1992 study) and since 1992 (the 1994 study). Global QL may be measured by a single item when brevity is desired [22]. To investigate global QL, the questionnaires dealt with the following items of which the first was used for investigating global QL by Nordbeck [23]:

1. "How do you feel about your life as a whole just now?" (PQL)
   - with five response categories: "very good, rather good, neither good nor bad, rather bad, very bad".
2. "Do you think that your life as a whole has become better or worse since you got your early retirement pension - compared with the year just before this pension?" (CQL 1992), "Do you think that your life as a whole has become better or worse compared with two years ago?" (CQL 1994)
   - with five response categories: "much better, better, unchanged, worse, much worse".

Other variables

In addition to questions concerning global QL, the questionnaires included questions concerning satisfaction with different aspects of life (home and family, dwelling, economy, work and leisure time), social network (quantity and quality), change in economic situation and loneliness since retirement, health problems (ailments/handicaps, symptoms, drug consumption, health care utilisation, change in health care utilisation, subjective health status and change in subjective health status), ADL status and change in ADL status, height and weight, smoking habits, change in smoking habits, previous work and working hours, unemployment experiences, view of disability pension as the best solution when granted and today, self-rated consequences of early retirement, positive self-image, immigration, civil status, spouse working or not, housing conditions today and five years ago, educational level. Information about the one or two diagnoses at retirement, work injury and year of retirement were given by the local social insurance office.

These variables were described and defined in previous articles (16, 19-20, 24).

Statistical analyses

A chi-squared test was used to compare proportions between groups. To decide the importance of subjective health for the QL among the DPs, Spearman's correlation coefficient (r_s) was used. The coefficient of determination (r_s^2) was calculated to estimate the rate (%) of the variance in QL explained by the association with the health variable.

Multivariate analyses were carried out by means of a logistic regression model with CQL as the dependent variable dichotomised by the median value 1992 into Improving (response categories "much better" and "better") and Unchanged or deteriorating (remaining response categories). Explanatory variables included in the model were those with a significant bivariate relation to the dependent variable (Table 10.2, 10.3). The level of significance was set to 0.05. In the logistic regression analyses, variables on an ordinal level or higher were dichotomised by using the median value to split the variable. The results from the logistic regression analyses were expressed as odds ratios (OR) with 95% confidence intervals (CI). Since OR is used to compute the relative risk ratio for a negative outcome, Table 10.2 and 10.3 show the OR for unchanged or deteriorating CQL.

The study was approved by the Committee on Ethics at the Faculty of Medicine, University of Lund (LU 289-91 and LU 327-93).

RESULTS

Self-rated improvement in QL since the year preceding the disability pension decision one to six years previously (CQL 1992) was significantly related to unchanged or improved subjective health and ADL status since retirement, satisfaction with social network in terms of close contacts outside the home, and satisfaction with leisure time activities and no difficulties in passing time (Table 10.2). Improvement in CQL 1992 was furthermore connected to: female gender, age >54 years, considering the disability pension to be the best solution today and having been out of work one year or more during the five years preceding the early retirement compared to no unemployment experience.

Female gender and unchanged or improved subjective health status from 1992 were furthermore significantly connected to a self-rated improvement in QL from 1992 to 1994 (CQL 1994, Table 10.3).

The number of years since retirement had no significant bivariate correlation to PQL or CQL in either 1992 or 1994. The correlation between CQL92 and CQL94 was $r_s=0.275$ (with a coefficient of determination, $r_s^2=0.076$).

The correlations (r_s) between PQL and subjective health status were 0.546 (1992) and 0.648 (1994). The coefficients of determination (r_s^2) were 0.298 (1992) and 0.420 (1994). Correlations (r_s) between CQL and change in subjective health status were 0.508 (1992) and 0.645 (1994) with coefficients of determination (r_s^2) 0.258 and 0.416 respectively.

DISCUSSION AND CONCLUSIONS

The importance of subjective health status for QL was demonstrated in our study. This is in line with results from population-based studies [12-15]. The rate of variance in PQL explained by the association with self-rated health status was 30% in 1992 and 42% in 1994. A similar very high connection was found considering CQL and change in subjective health status. The direction of the influence is not obvious, however, but presumably is mutual. It is hardly surprising that individuals with health problems, e.g.

DPs, who find that their health status has improved estimate their CQL to have improved as well. The importance of mobility as a predictor of positive perceived health has been established [25], and our study confirmed the importance of mobility as individuals who experienced decreasing restrictions in ADL found CQL to have improved irrespective of perceived health status (Table 10.2).

Table 10.2 Odds ratios (95% confidence intervals) for determinants of unchanged
or deteriorating QL since the year preceding the disability pension
(CQL) among DPs 1992 (n=376). Results from logistic regression
modelling of variables showing a bivariate correlation to CQL

Background variables *(n=347)*	OR	(CI)
Gender: male	**2.05**	(1.22-3.46)
Age <55	**1.93**	(1.12-3.30)
Immigrant	1.50	(0.77-2.95)
Early retirement not OK when retired	1.52	(0.66-3.50)
Early retirement not OK today	**4.74**	(1.002-22.39)
Out of work:		
(1 year	1.00	
< 1 year	4.83	(0.95-24.57)
not out of work	**2.86**	(1.05-7.79)
Health related variables *(n=318)*		
Self-rated health: bad	1.13	(0.64-2.00)
Deteriorating self-rated health	**2.32**	(1.27-4.25)
Medication	1.25	(0.44-3.53)
Medication: Analgesics	1.19	(0.52-2.76)
Health care not diminished	1.61	(0.94-2.75)
ADL deteriorating	**2.06**	(1.13-3.75)
Other variables *(n=289)*		
No close contacts outside the home	**2.57**	(1.05-6.30)
Leisure time: not content	**2.18**	(1.26-3.75)
Difficulties in passing time	**2.13**	(1.19-3.82)
Financial situation: not content	1.20	(0.69-2.11)
Financial situation deteriorating	1.68	(0.95-2.98)
Self-image not positive	0.90	(0.53-1.54)

ORs in bold when p<0.05

Variables not showing a bivariate relation to CQL and therefore not included in
the logistic regression model were: year of retirement, educational level, kind of work
before retirement, working hours, work injury, initiative to retirement, dwelling today,
dwelling five years ago, other ailments than musculoskeletal disorders (cardiovascular,
pulmonary, psychiatric, diabetes), number of symptoms, psycho-somatic symptoms,
neurotic symptoms, ADL status, smoking habits, change in smoking habits, BMI (body
mass index), diagnoses at retirement (see Methods for definitions), in hospital, visit to
caregivers (physician, nurse, physiotherapist, alternative care), medication
(cardiovascular, pulmonary, diabetes, sleeping pills), civil status, satisfaction with
home and family, number of contacts outside the home, loneliness, change in loneliness
since retirement, satisfaction with dwelling, satisfaction with previous job.

Table 10.3 Odds ratios (95% confidence intervals) for determinants of unchanged or deteriorating QL during the last two years (CQL) among DPs 1994 (n=352). Variables showing a bivariate correlation to CQL.

Background variables *(n=347)*	OR	(CI)
Gender: male	**2.21**	(1.09-4.48)
Immigrant	2.57	(0.97-6.80)
Work injury:		
no application	1.00	
not approved	1.35	(0.63-2.90)
approved	**1.99**	(1.04-3.83)
Health related variables *(n=324)*		
Self-rated health: bad	1.43	(0.59-3.48)
Deteriorating self-rated health	**2.98**	(1.26-7.07)
Medication	0.88	(0.31-2.47)
Medication: analgesics	2.01	(0.85-4.73)
Health consequences due to retirement: not positive	**4.45**	(2.10-9.40)
Other variables *(n=300)*		
Consequences due to retirement:		
contact with others: not positive	1.73	(0.78-3.80)
financial situation: not positive	0.91	(0.46-1.81)
leisure time: not positive	1.44	(0.71-2.91)
overall consequences: not positive	**4.49**	(1.73-11.66)

ORs in bold when $p<0.05$

Male DPs - compared to female DPs - more seldom considered their CQL to have improved since retirement (Table 10.2) and as a retiree (Table 10.3). Gender roles are known to vary with living conditions. In Sweden, there was a gender difference in the early 1990s concerning age and the importance of gainful employment and domestic work [26]. Men and women aged 51-60 showed the traditional gender roles: women considered domestic work as the basis for their self-esteem, and their identity was partly related to the professional occupation of their husbands, while the identity of men to a great extent was determined by their own job. Among younger women work was more important, and they appreciated the relative independence a job entailed. The traditional gender roles mean that for unemployed men, engaging in housework is considered as a substitute for what they really want: a paid job [27]. In our study most of the DPs belong to the ages characterised by traditional gender roles. Female DPs thus might appreciate retirement as a relief from the double responsibility of paid work and as primary domestic caregiver. The dissatisfaction among men forced to leave the labour market might affect CQL negatively among male DPs, since they have more difficulties in finding meaningful activities that could be seen as a substitute for gainful employment.

DPs with less than ten years left to old-age pension more often than the younger ones considered their CQL to have improved since retirement (Table 10.2). Paid work could be seen as purely instrumental, as a necessary evil for acquiring material resources, or as an activity of great positive value for the individual and society. According to the instrumental view, a disability pension should be a relief for everyone irrespective of age. But if work has value beyond the economic one, if work is to be seen as giving the individual a feeling of fulfilling important common social obligations, a disability pension must be harder to accept the younger you are. Our results support the latter view.

Our study shows that individuals who experienced a period of unemployment during the five years preceding the disability pension considered their CQL to have improved after early retirement more often than individuals who were not unemployed close to the early retirement (Table 10.2). Harmful effects of unemployment relative to self-esteem, self-confidence and psychic stress have been demonstrated in several studies [e.g., 27-28]. Life as an unemployed person and as a DP is similar with respect to the orientation of the individual towards meaningful activities other than paid work. Individuals suffering from musculoskeletal disorders, and with personal experiences of both unemployment and disability pension, obviously find life as a pensioner better.

The DPs who found the disability pension to be the best solution in 1992 considered their CQL to have improved more often than DPs who were not convinced that the retirement was justified (Table 10.2). This finding is in line with the results reported in a study concerning voluntary early retirement schemes [29], where the most important factor influencing QL among the retirees was the matching of expectations of further work at the point of decision. If a DP considers push factors to have had a strong influence on the decision concerning the disability pension, obviously the CQL, as well as PQL, is affected negatively.

To experience an improvement in CQL since the disability pension, the importance of a functioning social network and opportunities to indulge in leisure activities was stressed in our study (Table 10.2). Among old-age retirees, social and leisure activities are known to contribute substantially to QL [30]. A disability pension does not hamper the individual's motivation to participate in leisure time activities [10], and among DPs leisure activities might be stressed as compensation for the premature loss of a work role. If health problems restrict the opportunity of engaging in certain leisure activities highly valued by the individual and it is hard to find any other meaningful activities, QL must be affected negatively. Daily social contacts at work terminate when the person is granted a disability pension, and ill health might influence the potential to engage in social activities, thus reducing the social network among some of the DPs.

Gender and subjective health were the only variables related to change in QL since retirement, CQL 92, as well as at the follow-up, CQL 94 (Table 10.2 and 10.3). This indicates that it was the retirement itself (and the conditions leading to the retirement decision) that were of crucial importance for the change in CQL. Two years more as a DP did not influence most of the variables related to CQL 1992. This was confirmed by the finding that the number of years since retirement did not affect PQL or CQL in either 1992 or 1994. Obviously the CQL questions measured different aspects of change in QL. Only 8% of the variance in CQL 94 was explained by the association to CQL 92. The change in QL since retirement is thus something other than a change in QL as a disability pensioner.

ACKNOWLEDGEMENTS

This study was supported by grants from the National Swedish Social Insurance Board, Volvo Research Foundation and Volvo Educational Foundation, and the Kristianstad County Council, Sweden.

REFERENCES

1. Olsson-Frick, H.,1985, Early retirement in an international perspective II, The international study, *International Journal of Rehabilitation Research 8*, pp.161-80.
2. Guillemard, A-M,1991, International perspectives on early withdrawal from the labour force. In *States, labour markets, and the formation of old age policy*, edited by Myles, J. and Quadagno, J., (Philadelphia: Temple University Press), pp.209-26.
3. Kohli, M., 1994, Work and retirement: A comparative perspective, In: *Age and structural lag. Society's failure to provide meaningful opportunities in work, family and leisure*, edited by Riley, M.W., Kahn, R.L. and Foner, A., (New York: John Wiley & Sons), pp.80-106.
4. Guillemard, A-M and Rein, M., 1993, Comparative patterns of retirement. Recent trends in developed societies. *Annual review of sociology 19*, pp.469-503.
5. Stattin, M., 1997, Yrke, yrkesförändring och utslagning från arbetsmarknaden - en studie av relationen mellan förtidspension och arbetsmarknadsförändring. (Occupation, occupational change and exclusion from the labour market - a study of the relationship between disability pension and labour market change.), *Thesis 1998: 8.* Umeå, Sociologiska institutionen, Umeå universitet.
6. Schultz, K.S., Morton, K.R. and Weckerle, J.R. 1998, The influence of push and pull factors on voluntary and involuntary early retirees' retirement decision and adjustment. *Journal of Vocational Behavior 53*, pp.45-57.
7. Hetzler, A., 1994, Förtidspension under 1990-talet - myt och verklighet. (Early retirement during the 1990s - myth and reality) In *SOU 1994: 148. Förtidspension - en arbetsmarknadspolitisk ventil?* (Early retirement - a regulator of labour market policy?). Rapport från Sjuk- och arbetsskadeberedningen samt Arbetsmarknadspolitiska kommittén. Socialdepartementet, Stockholm, pp.21-31.
8. Staaf, L., Berglind, H. and Ekholm, J., 1995, Arbetsförmåga - ett nyckelbegrepp vid förtidspensionering. (Work ability - a key concept for early retirement) *Socialmedicinsk tidskrift 72*, pp.455-62.

9. Olofsson, G. and Petersson, J., 1995, Seven Swedish cases: Production regime, personnel policy and age structure in seven Swedish firms in the era of the Swedish model. Wissenschaftszentrum Berlin für Sozialforschung gGmbH (WZB), FS II, Berlin, pp.95-201.

10. Hedström, P., 1987, Disability pension: Welfare or misfortune? In *The Scandinavian model. Welfare states and welfare research*, edited by Erikson, R., Hansen, E.J., Ringen, S. and Uusitalo, H., (New York: ME Sharpe, Inc), pp.208-20.

11. Edén, L. and Ejlertsson, G. and Petersson, J., 1999, Quality of life among early retirees. *Experimental Aging Research 25*, pp.471-5.

12. Ventegodt, S.,1995, Quality of life in Denmark. Results from a population survey (Copenhagen: Forskningscentrets Forlag), pp.480-8.

13. Glatzer, W.,1987, Levels of satisfaction in life domains. *Social Indicators Research 19*, pp.32-8.

14. Bowling, A., 1995, What things are important in people's lives? *A survey of the public's judgements to inform scales of health related quality of life. Social Science & Medicine 41*, pp.1447-62.

15. Farquhar, M. ,1995, Elderly people's definitions of quality-of-life. *Social Science & Medicine 41*, pp.1439-46.

16. Edén, L. and Brokhöj, T., Ejlertsson, G., Leden, I. and Nordbeck, B., 1998, Is disability pension related to quality of life? *Scandinavian Journal of Social Welfare 7*, pp.300-9.

17. Ekerdt, D.J., Bosse, R. and LoCastro, J.S., 1983, Claims that retirement improves health. *Journal of Gerontology 38*, pp.231-6.

18. Salokangas, R.K.R. and Joukamaa, M., 1991, Physical and mental health changes in retirement age. *Psychotherapy and Psychosomatics 55*, pp.100-7.

19. Edén, L., Ejlertsson, G. and Leden, I., 1995, Health and health care utilization among early retirement pensioners with musculoskeletal disorders. *Scandinavian Journal of Primary Health Care 13*, pp.211-6.

20. Edén, L., Ejlertsson, G., Lamberger, B., Leden, I., Nordbeck, B. and Sundgren, P., 1994, Immigration and socio-economy as predictors of early retirement pensions. *Scandinavian Journal of Social Medicine 22*, pp.187-93.

21. Naess, S., 1989, The concept of quality of life. In *Assessing quality of life*, edited by Björk, S. and Vang, J., Linköping collaborating centre - LCC- Health service studies 1, 1989, pp.9-16.

22. Bowling, A., 1997, Measuring health. A review of quality of life measurement scales. (Buckingham: Open University Press).

23. Samuelsson, S.M., Bauer Alfredson, B. and Hagberg, B., Samuelsson, G., Nordbeck, B., Brun, A., Gustafson, L., and Risberg, J., 1997, The Swedish Centenarian study: a multidisciplinary study of five consecutive cohorts at the age of 100. *International Journal of Aging and Human Development 45*, pp.223-53.

24. Edén, L., Ejlertsson, G., Leden, I. and Nordbeck, B., 2000, High rates of psychosomatic and neurotic symptoms among disability pensioners with musculoskeletal disorders. *Journal of Musculoskeletal Pain 8*, pp.75-88.

25. Bryant, L.L., Beck, A., and Fairclough, D.L., 2000, Factors that contribute to positive perceived health in an older population. *Journal of Aging and Health 12*, pp.169-92.
26. Nordenmark, M., 1995, "Kvinnlig" och "manlig" arbetslöshet. Upplever kvinnor och män arbetslöshet på skilda sätt? ("Female" and "male" unemployment. Do women and men experience unemployment in different ways?). *Arbetsmarknad & Arbetsliv 1*, pp.31-43.
27. Glorieux, I., 1999, Paid work: a crucial link between individuals and society? Some conclusions on the meaning of work for social integration. In *Social exclusion in Europe: Problems and paradigms*, edited by Littlewood, P., Ashgate, Aldershot ,pp.62-81.
28. Oswald, A. J., 1995, Bitar i arbetslöshetspusslet. (Pieces of the unemployment puzzle). *Arbetsmarknad & Arbetsliv 1*, pp.5-29.
29. Maule, A.J.,Cliff, D.R. and Taylor, R., 1996, Early retirement decisions and how they affect later quality of life. *Ageing and Society 16*, pp.177-204.
30. Bernard, M. and Phillipson, C., 1995, Retirement and leisure. In *Handbook of communication and aging research*, edited by Nussbaum, J.F. and Coupland, J., (UK: Lawrence Erlbaum Associates Publishers), pp.285-311.

PART III
Age-conscious Personnel Policies and Productive Aging

11. The Role of the Psychosocial Environment in Promoting the Health and Performance of Older Workers

Amanda Griffiths

Institute of Work, Health & Organisations, University of Nottingham, UK

ABSTRACT

One economic implication of the ageing population is that workers may need to work longer and retire later than has been typical for several decades. However, much needs to be done to ensure that work remains as positive an experience as possible for workers throughout their career trajectories, and that it does not damage their health or productivity. Traditionally, much of the effort towards improving older workers' health has concentrated on improving the physical work environment or on health promotion. But we should not neglect to examine the effects of psychosocial factors (concerning the management and organisation of work) on psychological and physical health. This is the major focus of occupational health psychology. Much research to date in this area, notably that concerning stress, has produced 'age-free' models of the relationship between the psychosocial work environment and health. But this is based on assumptions that should be challenged. This paper focuses on the relationship between age and work-related health, and between age and productivity, with a particular emphasis on psychosocial factors. It concludes that there remains considerable scope for designing optimal work systems to harness the potential of older workers.

Keywords: *Age, Health, Management, Psychosocial factors, Stress*

INTRODUCTION

Many countries in the developed world are now facing a positive situation in which many of their citizens will live longer than ever before. Calculations within the European Union [1] suggest that between 2010 and 2030, substantial decreases in retirement ratios will be seen in all countries. For example, in 2010, for each retired pensioner there will be 3.3 people of working age in Germany, 4 in Finland, and 4.5 in Britain. By 2030 these figures will change to 2.5 (Germany), 2.5 (Finland) and 3.2 (Britain). A similar picture is revealed by calculating how many people of working age there will be for each dependent, where dependents include all those who are less than 19 years old and those who are over 60. With these figures, projections suggest that by 2025, ratios would be approaching 1 in many countries, implying that by 2025, each worker's productivity should ideally be sufficient to cover the living costs of another (non-working person) [1]. This represents a substantial change from current figures. It is clear that maintaining the health, skills and productivity of older workers will be important in maintaining many countries' competitive edge. In societies that aim to sustain a basic level of welfare for their entire populations, the economic implications of these demographic changes are considerable. Various solutions for reducing the possible financial deficits are technically possible but it has been suggested that one of the least unpopular options will be to keep workers at work for longer [2]. In addition to these financial arguments, it seems that some organisations are already experiencing skills shortages. Some are actively recruiting older workers (usually referring to those aged 45 and above) to compensate for these shortages.

Any increased involvement in work for the older members of society should be framed within informed and appropriate expectations, and should be productive, safe and healthy. It should not be unnecessarily stressful, nor have implications for health in retirement. Much needs to be done to protect against the 'empty prize' of longevity without quality of life [3]. So far, much attention has been focused on individual-level strategies: for example, improving work equipment or coaching healthier lifestyles. But there is now an increasing recognition of the design, organisation and management of work (psychosocial factors) as important risks to health.

Work can be a source of much satisfaction. It can provide purpose, satisfaction, meaning and challenge, as well as a vehicle for learning, creativity and growth. It affords opportunities to use skills, to demonstrate expertise, to exert control and to achieve success - many of the features offered by sport and leisure activities [4]. Many people currently report that work plays a significant part in their lives, providing psychological as well as material benefits. However, at the same time, it is known that the major contemporary challenges to health at work are those associated with the way work and work organisations are designed and managed [5-7]. The health problems currently reported by workers reflect this trend away from exposure to the traditional, physical hazards of work, towards those concerned with the way work is designed, managed and organised. It is clear that to ensure the quality of working life and to protect the health of workers as they grow older, close attention needs to be paid to the psychosocial environment.

In order to design work and work organisations to suit the needs of older workers, there are several important questions we need to address. What do we know about the health of workers as they age in relation to the psychosocial environment? Do we understand what aspects of work are stressful or satisfying for older workers? Are these different from those experienced by younger workers? How can we design work to be less stressful and more attractive to older workers and persuade them to remain in work? It is thought that many developed countries are facing similar issues, but in the remainder of this paper, data from Britain will be used in order to discuss these challenges.

AGE AND THE EXPERIENCE OF WORK-RELATED STRESS

In order to illustrate the possible relationship between age, health and work stress, three sets of figures will be used: (i) incapacity benefit data, (ii) surveys of work-related ill-health, and (iii) data on early ill-health retirements.

Incapacity benefit figures reveal how many people of working age (who have recently worked for a specified period) are unavailable for work because of ill-health. In 1997, the two most prevalent types of claim in Britain, with each category representing a quarter of all claims, were (i) mental and behavioural disorders including what could be described as 'stress-related' symptoms), and (ii) diseases of the musculoskeletal system and connective tissue [8]. These figures do not reveal whether these health problems are caused or made worse by work, but only that people who have recently worked suffer them and as a result are absent from work.

A stratified random sample of 40,000 people in Britain who were working or who had worked were asked whether they had suffered in the last 12 months from "any illness, disability or other physical problem" that they thought was caused or made worse by their work. Those who responded positively to this question were asked to contribute to a further survey where they were asked, about the nature of their illness, the job which they considered to have caused it (or made it worse), how many working days were lost (self-reported) and about their perceptions of certain aspects of their work [7]. This survey suggested that an estimated 19.5 million working days were lost to work-related illness in Britain in 1995. This represents an estimated 2 million people suffering from a work-related illness. Breaking down the data by age revealed that twice as many cases of psychological ill-health (stress, depression and anxiety) were found in older workers, (aged 45 to retirement age) than in younger workers. Very few cases were reported from the retirement age group, which suggests a reversible effect. A cohort effect is less likely explanation since the same trend was observed in the previous survey [9]. Furthermore, twice as many older workers reported physical conditions that they attributed (unprompted) to 'stress at work' (e.g. hypertension, heart disease, stroke or digestive disorders) than did younger workers. Where contacted, respondents' family doctors (general practitioners) agreed about the work-relatedness of these conditions.

A third source of information on the health of older workers is ill-health early retirements. According to the General Household Survey, the number of early ill-health retirements rose by 66% between 1972 and 1996 [10]. However, interpreting any trends in such data in relation to actual differences in the health of the working population is not straightforward. Part of any increase may be attributable to people's changing perceptions of health and to changes in corporate objectives and pension scheme policies and practices. However, some indication of the current picture is revealed by a survey of over 50,000 employees taking retirement from a cross-section of major employers. Of these, 14% retired early on the grounds of ill-health [11]. There are no centralised records in Britain about the cause of ill-health early retirements, but data from other countries such as the Netherlands and Sweden indicate that such retirements are increasingly made on the grounds of stress and musculo-skeletal disorders [12,13]. It is clear that the largest causes of work-related ill-health amongst the working population of Britain are stress and musculo-skeletal disorders. These are both known to be strongly associated with psychosocial factors - the way work is designed, organised and managed [14-16]. It is also apparent that older workers appear to be twice as vulnerable as younger groups [7]. It would clearly be useful to discover which particular aspects of the psychosocial working environment would are experienced as problematic or stressful by older workers.

PSYCHOSOCIAL HAZARDS FOR OLDER WORKERS

Occupational health psychologists have focussed on workers' own judgements of the way their work is designed, organised and managed and how those perceptions drive their emotions, their work-related and health-related behaviours and, ultimately, their health. Workers' appraisals of their working conditions, and the meanings they ascribe to them, are fundamental in understanding any relationship between those conditions and health outcomes [17,18]. Such self-reports have been found to predict health outcomes as well as or even better than independent measures [19-21].

Research has identified in very broad terms which characteristics of the psychosocial work environment can be stressful for most people. These largely concern difficulties with workload, work pace, working hours, organisational culture, participation and control, interpersonal relationships, career development, role-related issues and the home-work interface [17]. We may need to add other problems to this list - such as excessive working hours, lack of feedback, unsuitable or non-existent appraisal mechanisms, poor communication with senior management and inappropriate target-setting. However, most research into the relationship between work design, management and health has not explored 'age' as a variable in its own right. Age, when it has been considered, has been treated as a potential confound, has usually been partialled out statistically or simply ignored. It has consequently been assumed that what is 'bad' for one age group (in terms of the psychosocial environment) is 'bad' for all. The 'big picture' that is available from the scientific literature concerning harmful work characteristics may be masking important age-related differences. A common assumption has been that older and younger workers think about their work in similar ways, and make judgements on it in much the same way, and subsequently that any models based on a full age range will be meaningful.

However, recent research suggests that there are specific characteristics of work that are regarded as particularly problematic by older workers and therefore are likely to be particularly stressful for them. These problems are different from those reported by younger workers, even when doing the same job [22]. They largely reflect high-level contextual issues such as management systems and procedures and the knock-on effects of work on home life. Younger workers' concerns about their work appear more immediate and focused on task content. In similar vein, other research has found that older workers experience particular problems with lack of recognition, devaluating behaviours of supervisors and colleagues, and disappointment with management: all high level contextual issues [23].

The pioneering longitudinal studies conducted at the Finnish Institute of Occupational Health [24,1] have suggested that for older workers, certain psychosocial factors are significantly associated with decreasing 'work ability' - notably role conflict, fear of making mistakes or of failure, lack of influence over one's work, lack of professional development and lack of feedback and appreciation. These are all management issues. As well as predicting ill-health and sickness absence, such potential stressors have been found to be more powerful in the prediction of early retirement than physical problems [23].

In addition to such factors, it may be expected that the often-quoted discrimination against older workers (often, inaccurately, on the grounds of productivity) represents a source of stress. Discrimination against older workers has been common [25], although there are signs that more recently managers may have begun to acknowledge their commitment to work, experience, reliability, stability, and their problem-solving and interpersonal skills [26]. Much of this has been recognised in the scientific literature for some time [27,28]. Most reviews in the scientific literature report little consistent relationship between ageing and work performance [27-31]. Overall, older workers perform as well as younger workers. It has been proposed that despite decreases in certain cognitive and physical abilities, there is no observable decrease in older workers' overall performance, because what they lack in cognitive abilities they compensate for with an increase in job knowledge, skills [30] and various coping strategies such as better anticipation or more economical search strategies - all useful for problem-solving. However, despite these advances in research and in some managers' awareness, there is a considerable way to go in reducing discrimination and persuading many organisations of the potential of older workers.

CONCLUSIONS

Because older workers have certain known vulnerabilities, employers have an increased duty to take reasonable care for their health and safety. The psychosocial work environment should be designed to suit their abilities. For example, it should not require fast information processing, nor be very demanding on working memory. Particular recommendations for the development of the psychosocial working environment can be drawn from work in Finland [1]. For example, older workers need interesting work that involves life-long learning, opportunities to regulate their work themselves, acknowledgement and respect for their work, reduced or more flexible working hours, and opportunities for gradual retirement. Their supervisors need education and training about managing older workers, and knowledge of how to make use of their wisdom, work experience and interpersonal skills. Employers need a positive attitude towards the training of older workers, and should provide age-appropriate training.

Most so-called 'inevitable' age-related deteriorations may be variously countered by such changes, and by flexible and individual work designs, and support from well-informed management [32]. Employers would be well advised to capitalise on older workers' job knowledge, to make more use of them as mentors, to encourage horizontal as well as vertical mobility, and to allow greater flexibility. Age awareness programmes are important for all sections of the workforce, but especially for supervisors and managers. Above all, because many of the origins of stress for older workers may be local and context-specific, employers need to consult their own employees. The disadvantages of relying solely on a broad, context-free definitions of what is 'good' and 'bad' are becoming clear [33].

As yet, many traditionally managed organisations do not represent optimal systems for older workers and much remains to be done to instigate good practice [34]. Without supportive psychosocial environments and organisational cultures we cannot expect older workers to remain healthy, satisfied and productive at work, nor to outperform younger workers. Once attention has been paid to improving the psychosocial work environment, a healthier, less stressful and more productive future for older workers may emerge.

REFERENCES

1. Ilmarinen, J., 1999, *Ageing workers in the European Union: status and promotion of work ability, employability and employment*, Finnish Institute of Occupational Health, Helsinki.
2. Miles, D., 1997, Financial markets, ageing and social welfare. *Fiscal Studies*, 18, pp. 161-188.
3. World Health Organization, 1997, The World Health Report 1997. In *Conquering suffering, enriching humanity* (Geneva: World Health Organization).
4. Csikszentmihalyi, M., 1997, *Living well: the psychology of everyday life* (London: Weidenfeld & Nicholson).
5. Griffiths, A., 1998, The psychosocial work environment. In *The changing nature of occupational health*, edited by McCaig R.C. and Harrington M.J., (Sudbury: HSE Books).
6. Griffiths, A., 1998, Work-related ill-health in Great Britain. *Work & Stress*, **12**, pp. 1-5.
7. Jones, J.R., Hodgson, J.T., Clegg, T.A., and Elliott, R.C., 1998, *Self-reported work-related illness in 1995*, (Sudbury: HSE Books).
8. Social Security Statistics, 1997, (London: Sationary Office).
9. Hodgson, J.T., Jones, J.R., Elliott, R.C. and Osman, J., 1993, *Self-reported Work-related Illness*, (Sudbury: HSE Books).
10. Office for National Statistics, 1997, *Living in Britain-Preliminary Results from the 1996 General Household Survey*, (London: Office for National Statistics).
11. Income Data Services, 1998, *IDS Pensions Service Bulletin*, 115, May, (London: Income Data Services).
12. Goedhart, W.J.A., 1992, Ageing and the work environment. In *Ageing at Work: Proceedings of a European Colloquium*, Paris, June 1991, :European Foundation for the Improvement of Living and Working Conditions, Dublin.
13. Nygård, C.H., 1992, Ageing and work in Sweden. In *Ageing at Work: Proceedings of a European Colloquium*, Paris, June 1991, European Foundation for the Improvement of Living and Working Conditions, Dublin.
14. Bongers, P.M., deWinter, R.R., Kompier, M.A.J., and Hildebrandt, V.H., 1993, Psychosocial factors at work and musculoskeletal disease. *Scandinavian Journal of Work, Environment and Health*, **19**, pp. 297-312.

15. Moon, S.D. and Sauter, S.L., 1996, *Psychosocial Aspects of Musculoskeletal Disorders in Office Work*, (London: Taylor & Francis).
16. Sauter S.L. and Murphy, L.R., 1995, *Organizational risk factors for job stress*, (Washington DC: American Psychological Association).
17. Cox, T., Griffiths, A.J. and Rial González, E., 2000, *Research on work-related stress*, (Luxembourg: Office for Official Publications of the European Communities, European Agency for Safety & Health at Work).
18. Dewe, P., 1992, Applying the concept of appraisal to work stressors: some exploratory analyses. In *Human Relations*, **45**, pp. 143-164.
19. Stansfeld, S.A., North, F.M., White, I. and Marmot, M., 1995, Work characteristics and psychiatric disorder in civil servants in London. In *Journal of Epidemiology and Community Health*, **49**, pp. 48-53.
20. Bosma, H., Marmot, M.G., Hemingway, H., Nicholson, A.C., Brunner, E. and Stansfeld, S.A., 1997, Low job control and risk of coronary heart disease in Whitehall II (prospective cohort) study. In *British Medical Journal*, **314**, pp. 558-565.
21. Hedin, A.M., 1997, From different starting points-a longitudinal study of work and health among home care workers. In *European Journal of Public Health*, **7**, pp. 272-278.
22. Griffiths, A., 1999, Work design and management-the older worker. In *Experimental Ageing Research*, **25**, pp. 411-420.
23. Kloimüller, I., Karazman, R., and Geissler, H., 1997, How do stress impacts change with aging in the profession of bus drivers? Results from a questionnaire survey on 'health and competition' among bus drivers in a public transport system in 1996. In *From Experience to Innovation: Volume V, Proceedings of the 13th Triennial Congress, International Ergonomics Association*, edited by Seppälä, P., Luopajarvi, T., Nygård, C-H. and Mattila, M., Finnish Institute of Occupational Health, Helsinki, pp. 454-456.
24. Ilmarinen, J., Tuomi, K., Eskelinen, L., Nygård, C-H., Huuhtanen, P. and Klockars, M., 1991, Summary and recommendations of a project involving cross-sectional and follow-up studies on the aging worker in Finnish municipal occupations, 1981-1985. In *Scandinavian Journal of Work, Environment and Health*, **17** (Supplement 1), pp. 135-141.
25. Walker, A., 1993, *Age and attitudes*, (Brussels: European Commission).
26. Kodz, J., Kerseley, B. and Bates, P., 1999, *The fifties revival*, (Brighton: Institute of Employment Studies).
27. Rhodes, S.R., 1983, Age-related differences in work attitudes and behaviour: a review and conceptual analysis. In *Psychological Bulletin*, **93**, pp. 328-367.
28. Warr, P., 1994, Age and job performance. In *Work and aging: A European perspective*, edited by Snel, J. and Cremer, R., (London: Taylor & Francis), pp. 309-322.
29. McEvoy, G.M. and Cascio, W.F., 1989, Cumulative evidence on the relationship between employee age and job performance. In *Journal of Applied Psychology*, **74**, pp. 11-17.

30. Salthouse, T.A. and Maurer, T.J., 1996, Aging, job performance, and career development. In *Handbook of the psychology of aging*, edited by Birren J.E. and Schaie K.W., (San Diego: Academic Press), pp. 353-364.
31. Waldman, D.A. and Avolio, B.J., 1986, A meta-analysis of age differences in job performance. *Journal of Applied Psychology*, **71**, pp. 33-38.
32. Griffiths, A., 1997, Ageing, health and productivity-a challenge for the new millennium. *Work & Stress*, **11**, pp. 197-214.
33. Griffiths, A., 1999, Organizational interventions: facing the limits of the natural science paradigm. *Scandinavian Journal of Work, Environment and Health*, **25**, pp. 589-596.
34. Walker, A., 1999, *Managing an ageing workforce: A guide to good* practice, European Foundation for the Improvement of Living and Working Conditions, Dublin.

12. The Management of Work-related Stress with Regards to the Health of Older Workers

Tom Cox

Institute of Work, Health & Organisations, University of Nottingham, Nottingham, United Kingdom

ABSTRACT
Given the changing demography of work, it will become important in most developed countries to encourage older workers to remain at work longer and to ensure that working longer does not damage their health then or in later retirement. Surveys of the self-reported health of working and recently working people in Britain suggest that they stress is one of the major challenges that they and their organisations face. This paper considers the management of work-related stress in relation to the health of older workers, and describes the risk management approach developed and used in both private and public sector organisations in Britain and elsewhere by the Institute of Work, Health & Organisations.

Keywords: *Work-related stress, Worker health, Risk management, Older workers*

INTRODUCTION

The emotional experience of stress, in relation to work, is an outcome of the person's appraisal of the balance between their abilities and skills, on the one hand, and the demands of the work environment, on the other, taking into account control over work and support from others [*1]. The antecedents of work-related stress lay therefore, not only and not primarily in the person, but in the design and management of their work and of its social and organisational contexts. In many ways, the experience of stress at work indicates a failure of design and management systems.

With increasing age, the person's abilities and skills change, cognitive, physical and social, as do their cumulative experience and knowledge, and interests and motivations [2]. These changes can fundamentally alter the person's work ability and their appraisal of the stressfulness of their work. Interestingly, there is the suggestion that not all of these age-related changes are detrimental in terms of the experience of work-related stress [2]. Whatever, organisations have a continuing responsibility to design and manage the work of the older worker in such a way as to optimise its benefits and minimise the workers experience of stress. To achieve these twin goals, organisations will have to do two things: first, utilise methods of risk management that work for stress-related hazards, and, second, ensure that those methods are applied in a way that allows the analysis of age-related effects. It is particularly important to clearly delineate those stressors that are experienced by younger workers from those experienced by older workers.

The risk management approach initially focuses on the assessment of the design and management of the psychological, social and organisational work environments against likely stress-related health outcomes. The risk assessment data are translated into action programmes focused initially on reducing the risk of ill-health at source: primary prevention. However, it may also be necessary to supplement such interventions with additional protective measures for particular groups, such as training and enhanced support, and this may be the case for older workers. Interventions must be evaluated, with age of worker as a key variable, and the conclusions fed back to facilitate organisational learning, better management and design practice, and worker education and training.

DEVELOPMENT OF THE RISK MANAGEMENT APPROACH TO WORK-RELATED STRESS

The development of the risk management approach was initiated in the early 1990s with the publication of the author's report on "Stress Research and Stress Management: Putting Theory to Work" [1] and was funded largely by the Health & Safety Executive. The argument for its development lay in the changing nature of work and work problems and of work-related ill health. Examples can be taken from the European Surveys of Working Conditions [3,4] and the British Labour Force Surveys of the 1990s [5,6]. Together these clearly established that work-related stress and associated problems in the design and management of work were among the main challenges to occupational health in the 1990s not only in Britain but throughout the European Union. The interpretative framework implicit in the analysis of the stress-related data from the Labour Force Surveys was clearly that of the traditional health and safety equation of *HAZARD > RISK > HARM* with the additional suggestion that work-related stress might be a mediating factor in the relationship between hazard exposure and subsequent harm to health. It was argued by the author that if work-related stress is a health and safety issue, then it made sense to deal with it using the strategy that has been employed for managing other health and safety problems - the risk management paradigm [1,7]. Risk management is essentially an attempt at systematic and logical problem solving.

What of stress research as traditionally practiced ? It has been argued elsewhere that the traditional stress research paradigm is neither appropriate nor adequate as a starting point for risk management [1,7]. The main criticisms of the traditional stress research approach relate to the design, methodology and focus of the typical study particularly the emphasis on publishing comparisons across work groups allowing the construction of 'macro' models. Such an approach inevitably fails to capture the specific problems of particular work situations that need to be addressed to reduce risk. There has been a failure to allow for context and situation-specific effects. Furthermore, few research studies have selected or defined their samples or populations adequately enough for risk assessment purposes, and of the statistical analyses used in such studies have handled data at the level of the individual employee and not at the level of the work group. This reflects a clear bias towards dealing with work-related stress as an *individual* problem.

Traditional stress research mainly serves intellectual curiosity and career development while the risk assessment approach is designed and normally carried out to 'make a difference'. Where traditional stress research has informed practice it has been almost unavoidable that the "stress management" interventions it gives rise to are focussed on the individual and, as such, unacceptably imply an individual responsibility for the experience and management of work-related stress [7]. The risk management paradigm offers something very different in a methodological sense and is driven solely by the objective of reducing the risk to health associated with the experience of work-related stress.

GENERIC RISK MANAGEMENT MODELS

There are several substantive texts that together discuss not only the general principles of risk management and particular models [9,10,11] but also their scientific and socio-political contexts [12] All such models appear to incorporate or otherwise recognise five important principles or elements: [i] risk assessment is a crucial first step in risk management; [ii] it has to have a declared focus on a defined work population, workplace, set of operations or particular type of equipment (etc.); [iii] it logically informs subsequent risk reduction actions; [iv] those actions, in turn, have to be evaluated using a multiplicity of methodologies and measures; and [v] the entire process has to be actively managed. These features have been built into the approach to work-related stress developed by the Institute of Work, Health & Organisations.

RISK MANAGEMENT FOR WORK-RELATED STRESS

It is important to establish realistic expectations of what is achievable when adapting a generic model of risk management to work-related stress. Two issues are important: First, there cannot be an exact point-by-point translation of models developed for more tangible and physical risks to situations involving stress-related hazards. There is a need to think logically and creatively when adapting such models. The issues that arise should be decided in the light of (local) legal requirements and practical constraints, informed by our knowledge of applied science, and the decisions taken should be part of the overall evaluation of the process. Second, a risk management approach to work-related stress cannot be 'rocket science' in terms of its specifications, the absolute accuracy and specificity of its measures or the mechanisms underpinning its decision making. Nor does it have to be. The goal is a "*good enough*" system that provides a vehicle for progress in the improvement of working conditions and the reduction of stress, and that also facilitates compliance with current health and safety legislation. For example, United Kingdom legislation indicates that a risk assessment has to be 'suitable and sufficient' ([13], Regulation 3(1)) rather than perfect or the ideal. The assessment has to be good enough to identify major stress-related threats to worker health and does not have to be refined sufficiently to pick up all the different nuances of complaint that could exist in any work group. In the United Kingdom, decided cases and agreed out-of-court settlements [14] can be used to 'benchmark' this exercise. An assessment must be good enough to detect, at the very least, the sort of stress-related problems that are the subject of successful legal cases and settlements.

THE NOTTINGHAM APPROACH

At the heart of most risk management models are two distinct but intimately related cycles of activity: risk assessment and risk reduction. This is implicit in the European Commission's *Guidance on Risk Assessment at Work* [15]. The risk assessment and risk reduction cycles therefore form the basic building blocks for the four stage model described here, linked by the often underestimated processes involved in translating the output of one into the input to the other: translation. However, other components are specified in the model: see Figure 12.1.

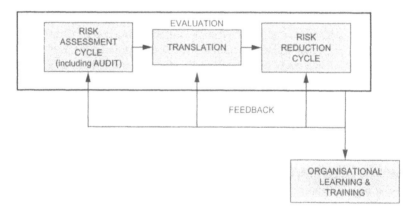

Figure 12.1 A Framework Model of Risk Management for Work-related stress

The model includes consideration of 'evaluation' and 'organisational learning and training'. Because all aspects of the risk management process should be evaluated - and not just the outcomes of the risk reduction stage - the 'evaluation' stage is treated as supra-ordinate to the other stages. The risk reduction stage, in practice, tends to involve not only prevention but also protective training and enhanced support, actions more orientated towards individual health and welfare. These may be of particular importance in relation to older workers working in a mixed age work group.

RISK ASSESSMENT FOR STRESS-RELATED HAZARDS

The initial stage, risk assessment, is designed to identify for a defined work group - with some certainty and in sufficient detail - significant sources of stress relating to the design and management of work, that can be shown to be associated with an impairment of the health of that group. The assessment should: [i] be carried out on a defined employee group, [ii] identify significant (non trivial) sources of stress related to their work, [iii] provide evidence of associated impairment to health, [iv] use reliable methods to ensure the certainty with which conclusions about risk can be drawn, and [v] work at a level of detail which can inform subsequent risk reduction activities. The logic underpinning risk assessment can be operationalised through a five-step system: see Figure 12.2.

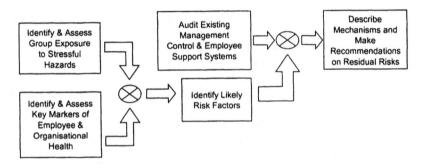

Figure 12.2 The Risk Assessment Strategy

[1] Hazard identification: reliably identify the stress-related hazards that exist in relation to the design and management of work, for specified groups of employees, and make an assessment of the degree of exposure. Since many of the problems that give rise to the experience of stress at work are chronic in nature, the proportion of employees reporting a particular aspect of work as stressful may be a "good enough" group exposure statistic.

[2] Assessment of harm: collect evidence that exposure to such stress-related hazards is associated with impaired health in the work group being assessed. The possible detrimental effects of work-related stress on a wide range of health-related outcomes should be considered, including symptoms of general malaise, behaviours such as smoking and drinking, and sickness absence.

[3] Identification of likely risk factors: explore the associations between exposure to stressors and measures of harm to identify "likely risk factors" at the group level, and to make some estimate of their size and significance. It is also necessary to be able to describe the possible mechanisms by which exposure to the stressors is associated with damage to the health of the assessment group or to the organisation.

[4] Audit existing management control and employee support systems: identify and assess all management systems that operate in relation to the control of stress-related problems and in relation to the provision of protective support for workers experiencing problems.

[5] Draw conclusions about residual risk: taking existing management control and worker support systems into account, make recommendations on the residual risk associated with the likely risk factors related to work-related stress: risk reduction.

The estimation of risk at the group level is effectively the estimation of risk for "the average employee", and can be contrasted with the estimation of risk for any specified individual. Where there is much difference between the two, then individual differences obviously exist. Some of these differences may themselves be age related.

There are two points to note here: First, the individual differences that exist can only operate through the person's interaction with their work environment. There is no other logical pathway by which their effects can be made manifest. Second, there is no evidence that the individual differences that exist in respect to the effects of stressors on health are any greater (or less) than those that exist in relation to other health hazards. The existence of individual differences does not negate the overall assessment exercise: rather, it adds an important extra dimension.

Particular emphasis is placed on the status of workers and their managers as "*experts*" and "*actors*" in relation to their own work and to changing that work. The collection of risk assessment data from them may therefore be cast as an exercise in knowledge elicitation and modelling. For this to work, managers and employees have to be involved in the process - *participate*. To do so effectively they have to be educated in the risk management paradigm and their expectations of its outcomes appropriately shaped. Interestingly, older workers may have different expectations of a risk management initiative for work-related stress than younger workers. This has proved the case in many of the Institute's projects.

Much of the data used in the risk assessment is based on the expert judgements of working people aggregated to the group level. Such judgements are sometimes labelled as 'subjective' and this is taken as grounds for dismissing the evidence so obtained. However, there is evidence to suggest that, at least for research into work-related stress, this is an appropriate and adequate strategy [16,17,18]. The work environment as experienced and judged may, in fact, be a better predictor of behaviour and health than the work environment as objectively measured. However, it is important to understand that referring to '*reported*' problems does not place an unquestioned value on those reports and judgements. The reliability, validity and accuracy of self-report and judgements are empirical questions, and, as such, can themselves be the subject of investigation. For example, social desirability effects (a common source of bias) can be tested for and screened out at several stages in the development of the assessment [18]. Respondents' patterns of reporting can be examined for halo effects: any evidence of differential effects would reduce the likelihood of the assessment data being driven by halo or similar effects (for example, negative affectivity [18]). The use of different measurement techniques and different sources of data (triangulation) should reduce the likelihood of common method variance [8,20].

STRESS-RELATED HAZARDS

Deciding what are and what are not stress-related hazards is far from straightforward. Moreover, it has often proved a difficult question in relation to the more tangible area of physical hazards. For example, Landy and colleagues [21] have asked whether the definition of stress-related hazards include aspects of work and organisations such as 'corporate policies, paid leaves of absence, promotion, health insurance coverage, etc.'. The answer must be an empirical one if the presence or absence of such things can be associated with harm to worker health then they are classed as a hazard. Stress-related hazards can often be conceptualised as part of a continuum that is represented by the hazard at one end and by the complementary health-enhancing factor at the other (for example, from *very low to very high* job control). Physical hazards, such as exposure to asbestos, often have a very different structure being negative per se without offering any obvious benefit for employee health. Other than not being present, they lack any potential health-enhancing role. This is an issue that requires both further theoretical exploration and empirical work.

It is possible that age-related differences exist in relation to exposure to stress-related hazards and in relation to the risk that they pose to health. With respect to exposure, age and length of service are normally strongly and positively correlated in most work groups, and in, at least, managerial and professional groups, also correlated with seniority and, if not, seniority then wage or salary. Essentially, the nature of work changes with length of service and age, and it is often difficult to find groups of different mean age doing similar work to, for example, explore similarities and differences in their experience of work and judgements on the adequacy of work design and management.

CONCLUSIONS

The complex aetiology of work-related stress represents a major challenge for applied science, and its mechanisms and causes may never be completely understood in their finest detail. However, there is a moral, as well as scientific and legal, imperative to act to reduce the harm to health caused by stress in the workplace. Risk management provides a framework for positive action -focused on prevention and on the organisation as the generator of risk- and has already proven successful in a number of varied occupational settings [8]. There is now a need to apply this methodology to managing the health of older workers. It is hoped that this article has provided a sound basis for progress in this area.

ACKNOWLEDGEMENTS

The author wishes to thank the Health and Safety Executive (Britain) for its support for this research and development work. The views expressed here are the authors' and do not necessarily reflect those of any other person or organisation.

REFERENCES

1. Cox, T., 1993, Stress Research and Stress Management. *Putting Theory to Work*, (Sudbury: HSE Books).
2. Griffiths, A., 1999a, Work design and management-the older worker. *Experimental Ageing Research*, 25, pp. 411-420.
3. European Foundation for the Improvement of Living and Working Conditions, 1992, *European Survey on the Work Environment*, Dublin, Ireland.
4. European Foundation for the Improvement of Living and Working Conditions, 1996, *Second European Survey on Working Conditions in the European Union*, Dublin, Ireland.
5. Hodgson, J.T., Jones, J.R., Elliott, R.C. and Osman, J., 1993, *Self-reported Work-related Illness*, (Sudbury: HSE Books).
6. Jones, J.R., Hodgson, J.T., Clegg, T.A., and Elliott, R.C., 1998, *Self-reported work-related illness in 1995*, (Sudbury: HSE Books).
7. Cox, T., Griffiths, AJ. and Rial-Gonzalez, E., 2000, *Research on Work-related Stress*, (Luxembourg: Office for Official Publications of the European Communities).
8. Cox, T., Griffiths, AJ., Barlow, CA., Randall. RJ., Thomson, LE. and Rial-Gonzalez, E., 2000, *Organisational Interventions for Work-related Stress*. (Sudbury: HSE Books).
9. Stranks, J., 1996, *The Law and Practice of Risk Assessment*. (London: Pitman).
10. Hurst, NW., 1998, Risk Assessment. *The Human Dimension*, (Cambridge: Royal Society of Chemistry).
11. Cox, S. and Tait, R., 1998, Safety. *Reliability and Risk Management*, (Oxford: Butterworth-Heinemann).
12. Bate, R., 1997, *What Risk ?*, (Oxford: Butterworth-Heinemann).
13. Health and Safety Commission, 1999, Management of Health and Safety at Work. *Management of Health and Safety at Work Regulations 1999, Approved Code of Practice and Guidance*, (Sudbury: HSE Books).
14. Griffiths, AJ., Cox, T. and Stokes, A., 1995, Work-related stress and the law: the current position. *Journal of Employment Law and Practice*, 2, pp. 93-96.
15. European Commission, 1996, *Guidance on Risk Assessment at Work*, (Brussels: European Commission).
16. Bosma, H., Marmot, MG., Hemingway, H., Nicholson, AC., Brunner, E. and Stansfeld, SA., 1997, Low job control and risk of coronary heart disease in Whitehall II (prospective cohort) study. *British Medical Journal*, 314, no. 7080.
17. Jex, SM. and Spector, PE. 1996, The impact of negative affectivity on stressor-strain relations: a replication and extension. *Work & Stress*, 10, pp. 36-45.
18. Spector, PE., 1987, Interactive effects of perceived control and job stressors on affective reactions and health outcomes for clerical workers. *Work & Stress*, 1, pp. 155-162.
19. Ferguson, E. and Cox, T., 1993, Exploratory factor analysis: A users' guide. *International Journal of Selection and Assessment*, 1, pp. 84-94.
20. Jick, TD., 1979, Mixing qualitative and quantitative methods: Triangulation in action. *Administrative Science Quarterly*, 24, pp. 602-611.

21. Landy, FJ., Quick JC. and Kasl, S., 1994, Work, stress and well-being. *International Journal of Stress Management*, **1**, pp. 33-73.

13. Post-Polio Fatigue and Aging: A New Problem in the Workplace in Japan

Satoru Saeki[1], Jin Takemura[2], Keinosuke Aridome[1],
Yasuyuki Matsushima[1], Hiromi Chisaka[1], Kenji Hachisuka[1]

[1] Department of Rehabilitation Medicine,
University of Occupational and Environmental Health, Kitakyushu, JAPAN
[2] Department of Rehabilitation Medicine, Cosmos Hospital, Oita, JAPAN

ABSTRACT

Post-polio syndrome (PPS) is generally defined as a clinical syndrome of new weakness, fatigue, and pain in individuals who have previously recovered from acute paralytic poliomyelitis. PPS occurs 30 or 40 years after an acute epidemic of poliomyelitis. It is believed that PPS is caused by distal degeneration of massively enlarged motor units from axonal sprouting after acute paralytic poliomyelitis. Clinically, overuse and aging are thought to be contributing factors. The purpose of this study was to examine the current symptoms related to PPS and working status in polio survivors who are of working age and living in Japan. A mailed questionnaire survey and a case study were performed. A total of 241 subjects were recruited, and their average age was 57 years. Half of the subjects complained of excessive fatigue or new muscle weakness, and three-quarters of them thought to have PPS. Approximately 90% of the subjects were working or ex-workers, and most of them had difficulty with ambulation. Particularly in the workplace, PPS individuals need special support from both rehabilitation medicine and occupational health services, including improved nutrition, achieving ideal body weight, regular and sensible exercise, frequent checkups and modifying working conditions.

Keywords: *Post-polio syndrome, Aging, Workplace*

INTRODUCTION

Poliomyelitis is no longer an epidemic disease in the western Pacific, including Japan, but in recent years a great many polio survivors experience late onset polio sequelae, called post-polio syndrome (PPS). PPS is generally defined as a clinical syndrome of new weakness, fatigue, and pain in individuals who have previously recovered from acute paralytic poliomyelitis [1,2]. PPS occurs 30 or 40 years after an acute epidemic of poliomyelitis. It is believed that PPS is caused by distal degeneration of massively enlarged motor units from axonal sprouting after acute paralytic poliomyelitis [1].

Clinically, overuse and aging are thought to be contributing factors [3]. This degenerative process can produce neuromuscular junction transmission defects, a possible cause of fatigue, and permanent denervation with resultant clinical weakness. Even though PPS is the most prevalent progressive motor neuron disease in the US today, Japanese PPS patients recently have begun to confront various problems. Since the last epidemic of acute poliomyelitis in Japan (1960s) occurred 10 years later than that in the US (1950s), issues relating to PPS also appeared approximately 10 years later in Japan (Figure 13.1).

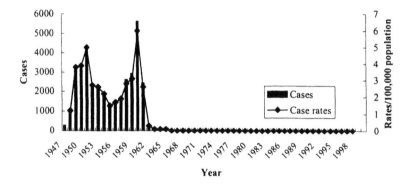

Figure 13.1 Cases and Rates of Acute Polio in Japan 1947-1998 [4]

Most new symptoms in PPS individuals relate to mobility, as paresis seems to be more common in the lower than in the upper extremities [5]. Since most patients have no or only limited disability with respect to personal activities in daily living (ADL) areas such as dressing and bathing, any new or increased degree of disability will be within instrumental ADL areas (e.g., cleaning, washing, shopping, transportation and working) and often relate to locomotive functions. The most common daily activities for which PPS individuals perceived increased difficulties were walking on level surfaces, climbing stairs, and meeting the physical demands of job and home. These functional limitations and disabilities resulted in handicaps in working conditions and reduced their quality of working life [6].

We present functional limitations and disabilities related to PPS in the Japanese working environment as observed from epidemiological research and a case study, and discuss special issues relating to working disability.

EPIDEMIOLOGICAL RESEARCH

We conducted a survey of polio survivors of working age who lived in the city of Kitakyushu (population approximately 1 million), Japan. A mailed questionnaire survey was performed at a cross-sectional setting.

The subjects were derived from the database of a registration of physically disabled persons' certificates in Japan. Based on the diagnoses of polio and/or polio-related disorders in the database, we selected polio survivors living in the city of Kitakyushu. We mailed letters of request to participate in the survey and questionnaire sheets to them, and 241 people ultimately were recruited as subjects for analysis.

The questionnaire consisted of the subject profile, impairment due to paralysis, physical condition, ADL, use of adaptive equipment and employment status. Whether or not the subjects experienced new weakness or fatigue after recovering from previous acute poliomyelitis, we asked them about the following eight physical conditions: fatigue, joint pain, numbness, walking, shortness of breath, muscle pain, muscle weakness and climbing stairs.

Their average age was 57 years. The majority of the subjects (63.5%) had had paralysis of one leg or arm only, whereas 5.4% stated paralysis in four or three extremities (Table 13.1). More than 60% of the study subjects used some kind of aid device; in particular, 20% of the subjects were using wheelchairs. Eighty-five percent of subjects reported new health problems: half of them complained of excessive fatigue or new muscle weakness. Three-quarters of subjects were thought to be PPS based on Halstead's diagnosing criteria for PPS [7].

Answers to questions about occupation reflect the person's capacity for work. Employment rates were high: 87% of subjects were employed since contracting polio, and white-collar occupations were predominant because of limited locomotive function (Table 13.2).

Table 13.1 Reported site of poliomyelitis paralysis

Paralysis	N (%)
Four extremities	7 (2.9)
Both arms and both legs	7 (2.9)
Three extremities	6 (2.5)
One arm and both legs	6 (2.5)
Both arms and one leg	0 (0)
Two extremities	73 (30.3)
Both legs	50 (20.8)
One arm and one leg	21 (8.7)
Both arms	2 (0.8)
One extremity	153 (63.5)
One leg	132 (54.8)
One arm	23 (9.5)

Table 13.2 Reported working conditions/employability

Working conditions	N (%)
Working/ex-workers	209 (87.1)
Full-time job	124 (59.9)
Part-time job	25 (12.1)
Self-employed	55 (26.6)
Others	3 (1.4)
Not working	32 (12.9)

CASE STUDY

A 44-year-old office clerk visited the post-polio clinic in 2000 with progressive weakness in his right arm and both legs, profound fatigue and pain in his right arm. At the age of 1 year 3 months, he experienced poliomyelitis with severe quadriplegia. While growing up, he recovered and regained the ability to walk with bilateral crutches and a Knee-Ankle-Foot Orthosis (KAFO) for his left leg. After graduating from high school at the age of 19, he was employed as a clerk in the Municipal office. He could walk a distance of one kilometer and climb five flights of stairs using the above adaptive orthotics.

In the late 1980s, when he was in his late 30s, he began to feel low back pain, weakness of right leg with muscle atrophy and difficulty in climbing stairs. In addition, he experienced cold intolerance, profound fatigue, gradually progressing muscle weakness and spinal scoliosis that affected his mental concentration and affected his job performance. He went to his office by car, moved on his wheelchair to his office and worked sedentarily at his workplace. Since his workplace passage was too narrow for his wheelchair to pass, he had to walk through the workplace using bilateral crutches and left KAFO.

Although he visited some doctors, no diagnosis was made. At the initial visit to our post-polio clinic 43 years after his acute illness, he had muscle weakness in his right arm and both legs and numbness in his right hand. The physical examination revealed moderate atrophy of both legs, mild focal atrophy of his right arm and asymmetric loss of strength in all four limbs. His trunk was moderately scoliotic to the left, and both feet were equinus. He had a leg length discrepancy, with the right leg 4 cm shorter than the left. The cranial nerves were normal; diminished deep tendon reflexes and no sensory losses were seen. Although he could walk with bilateral crutches and left KAFO on level ground for 50 meters, he usually used a wheelchair.

Laboratory data were unremarkable. The stress test using bicycle ergo meter revealed muscle damage due to overload (20W) as follows: creatine kinase (concentration of muscle enzyme in the blood, normal range: 62-287 IU/l) 155 IU/l at rest, 413 IU/l immediately after the test. Electromyographic studies (EMG) revealed giant motor units in four legs and acute denervation of right deltoid muscle due to right brachial plexus palsy. X-ray imaging showed severe spinal scoliosis to the left.

Based on his history, physical examinations, stress test and EMG findings, he was diagnosed with PPS. Also, he had right brachial plexus palsy, in other words compression or entrapment neuropathy of right axillary nerve by misuse of axillary crutches. This was because the increasing weight on the axillary crutch to support the scoliotic trunk compressed the above nerve against his right axilla. A comprehensive rehabilitation program and review of his KAFO and wheelchair were performed in an inpatient setting. His wheelchair was inappropriate and caused low back pain related to progressing scoliosis. Therefore, a new wheelchair that was adaptive and convenient for his workplace was made. After a four-week stay in our rehabilitation unit, he immediately returned to his former workplace. At that time, we advised him and his employer from the viewpoint of work-accommodation illustrated in Table 13.3. To avoid the overuse of leg muscles and excessive fatigue, we recommended a decrease in the workload and use of a wheelchair. Not all recommendations we presented were adopted due to financial constraints, but his working conditions have improved.

Table 13.3 Recommendations for return to work in the presented case

1. Securing of a wheelchair passage to the workplace from the parking lot, to avoid extra energy expenditure for ambulation.
2. Securing of the wheelchair passage in his workplace, for the above reason.
3. Working half time for the time being, to avoid excessive fatigue.
4. Job change to more sedentary work, to avoid overload and leg fatigue related to ambulation.

DISCUSSION

The estimated number of polio survivors is at least 640,000 in the US, 7,000-8,000 in Denmark and 50,000-100,000 in Japan [4,8]. Late sequelae of PPS include fatigue, muscle pain, joint pain, weakness of previously affected or unaffected muscles, cold intolerance and/or muscle atrophy [7,9,10]. These new problems may lead to loss of employment as well as new deficits in instrumental ADL, walking, climbing stairs, and personal assistance.

Our epidemiological data illustrated polio survival, symptoms, sociomedical and workplace adaptation in Japanese PPS. Evaluation of state of health can be described in at least three dimensions: 1) a social indicator: capacity of work, 2) an objective or professional measure, in this context meaning the need and current use of mobility aid devices, and 3) personal experience of state of health. Full-time work is the best indicator of a "good" health state without late onset polio sequelae [11].

A reduced capacity for work is an expected consequence of paralytic poliomyelitis. However, our result showed a high employment rate in polio survivors, and this finding agreed with another report by Halstead & Rossi, [9] showing that 93% of the sample had been working full-time or part-time since the onset of poliomyelitis. According to Bruno et al. [12], polio survivors are pushing themselves very hard, and disabled polio victims are often described as overachievers, demanding perfection in all aspects of their personal, professional, and social lives.

The case described here illustrated the problems of workplace disabilities and social handicaps in the Japan of today. As a result of the lack of recognition concerning PPS in Japanese society and among medical colleagues, the diagnosis of PPS has been delayed and PPS-related problems have been ignored. The social support system, including medical treatment, counseling and modifying working conditions, is necessary and should be legally established.

The case developed an entrapment neuropathy at the right axilla because of misuse of axillary crutches. These entrapment or compression neuropathies were, in a certain meaning, work-related disorders of PPS. Entrapment neuropathies seen in persons with paralytic polio are as follows [13]: median nerve at wrist (Carpal tunnel), ulnar nerve in hand (Guyon's canal) or at elbow (Cubital tunnel), brachial plexus (including axillary nerve). Application of appropriate rehabilitation, use of suitable orthotics and advice in the workplace can prevent the above entrapment neuropathies.

Medical rehabilitation was effective for the case documented here. The effectiveness of rehabilitation medicine for PPS has been established [10,13]. It has been demonstrated that people with a history of polio could improve muscular strength and endurance as well as cardiovascular conditioning by way of an exercise program [14].

Agre stated the principle of therapeutic exercise and recommendations as follows [10]: 1) weight loss, if necessary; 2) rest, or avoidance of overuse of damaged muscles and excessive fatigue; 3) appropriate gentle exercise when disuse is suspected; 4) muscle strengthening exercise by low-load and high frequency principle; 5) appropriate adaptive equipment such as orthotics, canes and wheelchairs, preventing overuse of damaged muscles, and 6) correction of a leg length discrepancy and ambulation, to reduce energy expenditure. This principle may help PPS individuals to better control symptoms of fatigue, and is further reinforced by our clinical experiences in workplace accommodation.

Frequently, polio survivors have managed to move through life at a great cost in energy and by performing a kind of balancing act: using preserved muscles to the maximum and often in ways that were never intended [15]. Their aging has the effect of disrupting that balance and exacting a price far greater than when it occurs in an intact person. In turn, the underlying disability appears to accelerate and possibly magnifies the usual effects of aging. Strength is deteriorating among PPS at a rate higher than that associated with normal aging [16]. This deterioration is not occurring in the extensor, or so-called "weight-bearing" muscles, but is occurring in many of the upper-extremity muscle groups and in the flexor muscles in the lower-extremities. Problems of aging such as the illustrated muscle weakness are major concerns for PPS in workplace accommodation.

CONCLUSIONS

We describe the epidemiological data of polio survivors and a case of PPS with working disabilities. In Japan, the exact number of polio survivors is unclear due to lack of a nationwide survey; however, the majority of polio survivors face new health problems from PPS. Particularly in the workplace, PPS individuals need special support from both rehabilitation medicine and occupational health services, including improved nutrition, achieving ideal body weight, regular and sensible exercise, frequent checkups and modifying working conditions.

REFERENCES

1. Cashman, NR., Maselli, R., Wollmann, RL., Roos, R., Simon, R. and Antel, JP., 1987, Late denervation in patients with antecedent paralytic poliomyelitis. In *N Engl J Med 317*, pp. 7-12.
2. Dalakas, MC., Elder, G., Hallett, M., Ravits, J., Baker, M., Papadopoulos, N., Albrecht, P. and Sever, J., 1986, A long-term follow-up study of patients with post-poliomyelitis neuromuscular symptoms. In *N Engl J Med 314*, pp. 959-963.
3. Trojan, DA., Cashman, NR., Shapiro, S., Tansey, CM. and Esdaile, JM., 1994, Predictive factors for post-poliomyelitis syndrome. In *Arch Phys Med Rehabil 75*, pp. 770-777.
4. Statistics and information department, minister's secretariat, ministry of health and welfare, Japan. (Japan: Statistics on communicable disease), 1998.
5. Grimby, G., 1995, Functional limitations and disability in post-polio. In *Post-polio syndrome*, edited by Halstead, LS. and Grimby, G., (Philadelphia: Hanley & Belfus, Inc), pp. 165-175.
6. Saeki, S., Takemura, J., Matsushima, Y., Chisaka, H. and Hachisuka, K., 2001, Workplace disability management in post-polio syndrome. In *J of Occupational Rehabilitation*, **11**, pp. 299-307.
7. Halstead, LS. and Rossi, CD., 1987, Post-polio syndrome: clinical experience with 132 consecutive outpatients. In *Research and clinical aspects of the late effects of poliomyelitis*, edited by Halstead, LS. and Wiechers, DO., (White Plains, NY: March of Dimes Birth Defects Foundation), pp. 27-38.
8. Lønnberg, F., 1993, Late onset polio Sequelae in Denmark. In *Scand J Rehab Med Suppl 28*, pp. 7-15.
9. Halstead, LS. and Rossi, CD., 1985, New problems in old polio patients: results of a survey of 539 polio survivors. In *Orthopedics 8*, pp. 845-850.
10. Agre, JC., 1995, Local muscle and total body fatigue. In *Post-polio syndrome*, edited by Halstead, LS. and Grimby, G., (Philadelphia: Hanley & Belfus, Inc.), pp. 35-67.
11. Lønnberg, F., 1993, Late onset polio sequelae in Denmark. In *Scand J Rehab Med Suppl 28*, pp. 24-31.
12. Bruno, RL. and Frick, NM., 1991, Psychology of polio as prelude to post-polio sequelae: behavior modification and psychotherapy. In *Orthopedics* **14**, pp. 1185-1192.

13. Windebank, AJ., 1995, Differential diagnosis and prognosis. In *Post-polio syndrome*, edited by Halstead, LS. and Grimby, G., (Philadelphia: Hanley & Belfus, Inc.), pp. 69-88.

14. Gawne, AC., 1995, Strategies for exercise prescription in post-polio patients. In *Post-polio syndrome*, edited by Halstead, LS. and Grimby, G., (Philadelphia: Hanley & Belfus, Inc.), pp. 141-164.

15. Halstead, LS., 1995, The lessons and legacies of polio. In *Post-polio syndrome*, edited by Halstead, LS. and Grimby, G., (Philadelphia: Hanley & Belfus, Inc.), pp. 199-214.

16. Klein, MG., Whyte, J., Keenan, MA., Esquenazi, A. and Polansky, M., 2000, Changes in strength over time among polio survivors. In *Arch Phys Med Rehabil 81*, pp. 1059-1064.

14. Work Situation Evaluation as a Prerequisite for Productive Aging of Engineers and Innovators

Klaus-Dieter Fröhner

Institute of Ergonomics, Technical University of Hamburg-Harburg
Hamburg, Germany

ABSTRACT

Engineers and innovators developing new products are playing an important role in highly industrialized nations. The purpose of the study was to understand special problems of aging that arise in German companies in the small segment of the work force consisting of engineers and innovators working in research and development. As the output of the educational system is very small, the shortage of engineers is dramatic, thus there is a need for older engineers to stay longer in the work force. Therefore, the variety of tasks, the activities across their domains and their attitudes towards organization were examined, mainly via a questionnaire, thus giving findings for organizational layout. 37 engineers were questioned in 25 companies. The group of engineers and innovators can be viewed as a predecessor of what can happen in small segments of the labour market with very skilled employees working under high mental demand.

Keywords: *Shortage of engineers, Work situation evaluation, Research and development, Engineers and innovators*

INTRODUCTION

In the approach to overcoming or at least reducing the difficulties that accompany aging and demographic change, the need to focus on this problem within the upcoming decades is being gradually recognized. One reason for the gradual recognition of the problem is seen in the declining participation rates of older workers over many years. This meant that the problem of integrating older workers in the work force seemed to lie far ahead.

That perception of the problem supported the emphasis in research areas such as the structure of the work force of aging societies [1], physiological and psychological problems [2,3] and age-associated diseases and their intervention [4]. Wider approaches towards the problem are the work ability index [5] and nation-wide programs for the older work force such as those in Finland [6] and Japan [7,8]. The knowledge accumulated here should form a solid basis for determining what can and should be focused on when older workers should stay longer in the work force.

The situation described here is not going to arise in the coming years. It is already happening today, but only for a small segment of the German work force, namely engineers and innovators developing new products. Engineers and natural scientists working in design as well as in research and development belong to this group, which can be viewed as a predecessor of what can happen in small segments with very skilled employees working under high mental demand. It occurred earlier in this segment of the work force than in the German labour market as a whole because two developments coincided: demographic change reduced the number of younger people and the restructuring of German industry due to globalization and reunification had a strong impact. As a result of the restructuring, many older engineers lost their jobs and remained unemployed. Therefore, the number of students studying mechanical and electric/electronic engineering science dropped to 30% of the previous level in the mid 90s. Consequently, the output of the educational system today is low. On the other hand, the demand for skilled engineers who are well acquainted with new technologies has been very high for some time [9]. One option for reducing the shortage of engineers can be seen in the possibility of having older engineers and innovators stay longer in the work force and reducing their desire for an early exit.

The approach of generating knowledge in order to provide some assistance should be the first step in understanding this special problem of aging and demographic change. It is necessary to look at the special working situation of engineers and innovators: how they work, what their differentiated tasks are, what their attitudes are towards management issues, etc. In this context, the questions are: how is the work designed and how are the management guidelines set up, when looking at it from a wider viewpoint. This viewpoint has been considered as an under-researched area [10].

SUBJECTS AND METHODOLOGICAL APPROACH

Engineers and innovators developing new products are playing an important role in highly industrialized nations as well as in the companies where they work. The consequence is an exposed position in the companies, which is often bolstered by patents they own. Because of this, they sometimes do not feel closely linked to the management schemes being employed in the companies. This complicated situation made a field study as a first step much more likely to gain results than any laboratory set-up.

The idea was not to look at the work performance of the engineers and innovators, because that would be of little success due to the level of the tasks they perform in different companies. The variety of tasks, the activities across their domains and their attitudes towards the organization were the focus.

The assumption was that the arrangements of tasks we find in the field of study will be an implicit match of personal desire and management strategy due to the exposed position of innovators. Personal desire is a very strong factor in how work tasks are arranged, because over the years engineers and innovators developing new products gain a high competence, especially when dealing with cross-sectional application of technologies. Thus, management is likely to follow their wishes. Furthermore, the organization in research and development as well as in design normally, especially in medium-sized companies, is not shaped as much by rigid organizational rules as at shop floor level (Figure 14.1).

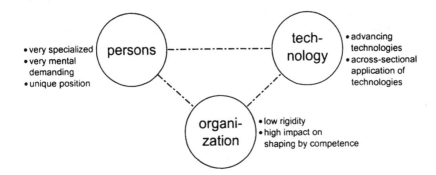

Figure 14.1 Characterizing dimensions of work situation evaluation of engineers and innovators developing new products in the examined companies

The studies were carried out in 25 small, medium-sized and large German companies, mainly belonging to the mechanical and electrical/electronic sector, so as to cover a wide range of what is realized in actual work in industry. They were part of a qualitative study that looked for innovation and performance in companies' work forces in general [11]. The studies of the engineers and innovators were carried out via a semi-structured questionnaire. The context of the work situation was also supported by document analysis for the production program and estimates concerning the applied technologies.

WORK SITUATION EVALUATION

With 37 innovators investigated, aged between 26 and 62, one to three engineers and innovators were looked at in the 25 companies. As excerpts of the work situation [12] three empirical findings are depicted. To clarify the different viewpoints according to age, the innovators were split into groups almost equal in number. The formulation of groups was influenced by the judgement in the electronic industry that an innovator aged over 40 years is seen as old.

The tasks carried out most often by the engineers and innovators questioned (Figure 14.2) show that design, development and production engineering are the core activities. Design is seen as a creative process, generating new products out of elements and parts that did not exist before. Development is seen as the process of converting creative ideas. These core activities are seen as the dominant activities by innovators older than 39 years. The investigation allows the conclusion that innovators older than 39 realize a larger variety of tasks than younger innovators. The tasks realized often show that the domains of the innovators are very wide, integrating design, production and marketing.

The main sources of stimulus for the innovation process are literature, customers, production and colleagues for innovators older than 39 years (Figure 14.3). The same tendency as depicted before can be drawn from the sources of stimulus: the older innovators get greater stimulus from a larger variety of sources than the younger ones. The stimulus is high even from sources that companies see as restrictions: production, compulsions and competitors. Management techniques such as kaizen and creative techniques are insignificant.

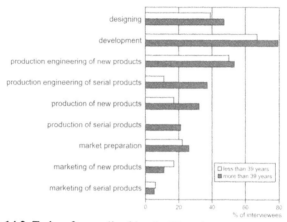

Figure 14.2 Tasks often realized by the 37 engineers and innovators studied

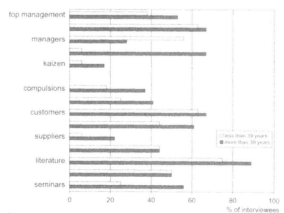

Figure 14.3 Stimulation for the innovation process of engineers and innovators studied [n=37]

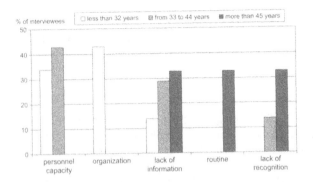

Figure 14.4 Barriers to innovation (n=37) perceived by innovators

A very good hint to how the organization is perceived by the engineers and innovators can be drawn from the question concerning the barriers to innovation (Figure 14.4). Here the innovators were split into three groups (young (up to 32 years), middle-aged (33 to 44 years), and old (over 45 years)).

Only lack of information was seen in part by all groups as a barrier. The higher time pressure felt by younger and middle-aged innovators can be seen in that they named personnel capacity and organization as barriers. Routine and lack of recognition were the main barriers for older innovators. This shows that the groups must be cared for in different ways.

FINDINGS AND INTEGRATION

The findings show that a very differentiated approach must be realized in order to develop work design and management issues so that older innovators not only remain productive, but so that younger and middle-aged innovators find a satisfying situation:

- the shaping of tasks should take into consideration life span models as generally proposed by [10]
- the variety of tasks must be adapted for the increasing ability to overview complex processes
- the creative part of the design process should not be reduced, but rather should be enlarged to create stimuli to overcome routine and realize productive aging
- the reduced acceptance of organizational rules should stimulate reorganization of research and development
- the difference in the accentuation of the work situation should lead to seeing ambiguity as a guideline, instead of rigidity.

The solution for stimulating this process in many companies is seen in installing the findings and the case studies on a structured CD-ROM together with some basics about aging, so that innovators and their managers can find out for themselves what can be of value for their particular work situation.

ACKNOWLEDGEMENTS
This study is a part of a series of research project on demographic change and the
future of labour, supported by the Federal German Ministry of Education and Research.

REFERENCES
1. Ilmarinen, J., 1999, Aging workers in the European Union-Status and promotion
    of work ability, employability and employment. Finnish Institute of Occupational
    Health, Ministry of Social Affairs and Health, Ministry of Labour, Helsinki, pp.
    274.
2. Shephard, R.J., 1999, Age and Physical Work Capacity. In *Experimental Aging
    Research*, **25**, pp. 331-343.
3. Tuomi, K., Seitsamo, J. and Huuthanen, P., 1999, Stress Management, Aging,
    and Disease. In *Experimental Aging Research*, **25**, pp. 353-358.
4. Bezooijen, van C.F.A., 1996, Biological aging, theories of aging, age-associated
    diseases and their intervention. In *Aging and Work 3*, edited by Goedhard, W.J.A.,
    ICOH Scientific Committee "Aging and Work", pp. 19-35.
5. Tuomi, K., Ilmarinen, J., Jahkola, A., Katajarinne, L. and Tulkki, A., 1998, Work
    Ability Index, Finnish Institute of Occupational Health, Helsinki, p. 34.
6. Ilmarinen, J. and Louhevara, V., 1999, Finn Age-Respect for the Ageing: Action
    programs to promote health, workability and well-being of ageing workers in
    1990-1996. *People and Work, Research Reports 26*, Finnish Institute of
    Occupational Health, Helsinki, p. 308.
7. Kumashiro, M., 2000, "Kaizen" activities at workplace for older workers in
    Japanese companies. In *Aging and Work 4*, edited by Goedhard, W.J.A., Ilmarinen,
    J., Kumashiro, M., Marklund, S. and Volkoff, S., ICOH Scientific Committee,
    "Aging and Work", pp. 101-108.
8. Muto, T., Hsieh, S.D., Sakurai, Y., Sawada S. and Sugisawa, H., 2000, Evaluation
    of workplace health promotion programs for older workers in Japan. In *Aging
    and Work 4*, edited by Goedhard, W.J.A., Ilmarinen, J., Kumashiro, M., Marklund,
    S. and Volkoff, S., ICOH Scientific Committee, "Aging and Work", pp. 82-93.
9. VDMA, 1998, Wie viele und welche Ingenieure benötigt der Maschinenbau in
    Zukunft? Maschinenbau konkret, Verband Deutscher Maschinen-und Anlagenbau
    e.V., Frankfurt, p. 6.
10. Griffiths, A., 1999, Work Design and Management-The Older Worker. In
    Experimental Aging Research, 25, pp. 411-420.
11. Köchling, A., Astor, M., Fröhner, K.-D., Hartmann, E.A., Hitzblech, T., Jasper,
    G. and Reindl, J., 2000, Innovation und Leistung mit älterwerdenden
    Belegschaften, Rainer Hampp Verlag, München und Mering, p. 390.
12. Fröhner, K.D. and Nawroth, K., 2000, Im Zentrum der betrieblichen Innovation:
    Jüngere und ältere Innovierer in Produktentwicklung und Konstruktion. Ju:
    Innovation und Leistung mit älterwerdenden Belegschaften,edited by Köchling,
    A., Astor, M., Fröhner, K.-D., Hartmann, E.A., Hitzblech, T., Jasper, G. and Reindl,
    J., 2000, Rainer Hampp Verlag, München und Mering, pp. 221-264.

15. Intergenerational Relations at Work in Sweden and the UK

Ingrid Johansson

*Department of Social Sciences, Mälardalen University,
Eskilstuna, Sweden*

ABSTRACT

In most European countries today, people risk being considered old in their working lives already at the age of 40, 45, or 50. Younger people are in demand and older people seem to have a decreasing value. An aim of the study was to extend the understanding and knowledge of this phenomenon by studying younger people's perceptions and experiences of older colleagues and managers in Sweden and the UK. Another aim was to collect examples of practical measures that could be applied in workplace to encourage intergenerational co-operation. The results were based on semi-structured interviews with 32 people, 16 in each country, aged 25-34. The analysis was accomplished manually and using a computer package for qualitative analysis. The results showed that younger people appreciated older people's experience, but found it easier to communicate and socialize with people of their own age. Negative stereotypes of older people could have an effect on younger people, who felt pressured not to waste time and to make a career quickly, before it was too late. Mentoring, team working in cross-age groups, and social activities outside the workplace were mentioned as examples of practical measures for the encouragement of intergenerational co-operation.

Keywords: *Co-operation, Older workers, Younger workers*

INTRODUCTION

Research shows that older workers are at a disadvantage on account of their age in the European labour market [1-5]. According to the researchers, with the increase of unemployment, older people have been expected to make way for younger people and have been persuaded or forced into early retirement. The early exit policy also had consequences for those older workers who remained in the labour market. Workers were considered too old at increasingly lower ages by employers and sometimes also by themselves. The existence of age barriers in recruitment is confirmed by a study of job advertisements in Sweden and Germany [3]. The results showed that, when age requirements were used, 25-35 was the most frequently mentioned age bracket in the two countries, whereas ages over 40 were rare.

Although age barriers are particularly visible and obvious in job recruitment and early retirement, they exist in other situations as well. According to the results of an interview study with middle-aged people in Sweden [3], age was perceived to be of significance in recruitment, early retirement, training, development, promotion, demotion, change of branch and profession, work abroad, and financial rewards.

The results of this study also showed that reasons for employers' preferences for younger people were attributed both to the older employees and to the employers. Examples of reasons ascribed to the employees were their supposed reluctance to change, greater need for security, and attitude of superiority. They might also be considered negative, tired, and less flexible and mouldable than younger people. Examples of reasons ascribed to the employers were new technology, lack of flexibility in the organisation, younger managers' fear of competition, prejudices, ageism, age discrimination, and conflicts between generations [3].

Intergenerational relations are a topic of many research projects. There are, for example, studies on changing family structures, the role of grandparents, adult children's care for their aging parents, intergenerational learning programmes, and socio-economic policies and their consequences [6]. However, not much research has been done on intergenerational relations in working life, and even less has been done as seen from the perspective of younger people. This is surprising, considering how influential younger people seem to be at work.

During academic year 1999/2000, an interview study with people aged 25-34 in Sweden and the UK was carried out. An aim of this article is to present some of the results of this study, focusing on intergenerational relations at work. Another aim is to present examples of practical measures mentioned by the interviewees, which could encourage intergenerational co-operation.

MATERIAL AND METHODS

Interviews were conducted with 16 younger people in Sweden and 16 in the UK, i.e., in total 32 interviews were carried out. The interviewees were all between 25 and 34 years of age, had a university education, and experience of working with age 45+ colleagues and managers. All of the British interviewees lived in Sheffield and all of the Swedish interviewees in Stockholm. The sample strategy was non-random and purposive and the goal was to find information-rich interviewees.

There was an equal number of male and female interviewees, i.e., eight younger men and eight younger women in each country. Many different kinds of organisations of varying sizes and branches were represented. In the private sector there were, for example, interviewees from industrial companies, banks, financial agencies, wholesalers, retailers, and IT companies. Examples of organisations from the public sector were government departments, local authorities, schools and universities. Due to the sampling method, whereby in most cases individuals were selected and not organisations, the choice of organisations is not claimed to be representative.

All the Swedish interviews were carried out before the British interviews. Some of them were conducted in 1998, whereas most of them and all of the British interviews were conducted in 1999. The interviews took between 45 and 90 minutes with an average of about 70 minutes.

The interviews were semi-structured and an interview guide was constructed and used. This was used in an open-ended way, i.e., with many follow-up questions that varied from interview to interview. All of the interviews were tape-recorded and the data collected was transcribed word for word. The Swedish material was also translated into English. Then the material was coded and arranged into preliminary categories and sub-categories inspired by grounded theory [7]. During the analysing process the data was scrutinised for patterns, themes, and regularities as well as contrasts and irregularities [8]. Special focus was placed on deviant and contradictory statements.

As a complement to the manual sorting and analysis processes, and as a way of using data analysis triangulation, all interview texts were imported into the qualitative analysis computer package QSR Nud*ist Vivo (Nvivo) and the material was coded. Some of the codes from the manual coding were used with some new codes being added. Triangulation was used both to increase the possibilities of finding new categories, subcategories and links between the categories and thus complete the manual analyses, and in order to verify the results of the manual analysis and increase the trustworthiness and credibility of the results.

Traditional ethical concerns in social sciences include informal consent, the right to privacy and protection from physical and emotional harm [9]. These and other ethical criteria compiled in "The ethical recommendations of The Swedish Council for Research in the Humanities and Social Sciences" [10] have been followed. Thus, the interviewees were informed of the aims of the study, that participation was voluntary and could at any time be interrupted, that the material would be treated in a confidential way and that it would not be used for other purposes than research. Quotations that might reveal the identity of an interviewee have been avoided.

RESULTS

The results have been organised into seven sections beginning with "Stereotypes, prejudices, and attitudes" and "Effects of prejudices against older people on younger people". Then follow sections on "Younger and older managers", "Intergenerational communication", "Women and men", and "Younger people's parents". Finally, there is a section on "Practical measures for the encouragement of intergenerational co-operation".

Stereotypes, prejudices, and attitudes

Generally, the interviewees found many advantages of older colleagues and managers. Older people's professional experiences as well as life experiences were appreciated. Having experience also meant having knowledge. The younger people could learn by watching older people or by asking them questions. Older people were considered to have a more relaxed approach, were calmer and more patient. They knew what was important and what was not, could refer to similar situations and thought things through. Furthermore, older people tended to be confident in their ability and they inspired respect.

However, the younger interviewees also mentioned quite a few perceived disadvantages of older colleagues and managers as well as stereotypes and prejudices. For example, older people were expected to be reluctant to change, and not open to new ideas and new technology. Not everybody agreed, though, that reluctance to change increased with age. Some pointed out that it might have to do with personality or with the length of time in the organisation. A Swedish man said:

> People who have stayed very long in the same workplace are probably more reluctant to change. Otherwise they would have changed jobs. I guess there is a correlation.

Older people were also supposed to be slow learners and less motivated to learn than younger people. The proverb "You can't teach an old dog new tricks" was frequently mentioned. A British young man summarised some stereotypes and prejudices against older people:

> People would say they are dinosaurs. They don't have a clue what is going on. They are old and stick in the mud. They just stick there. They won't change. They won't do anything. A general view is that they won't change, I think, and they are not in touch with the society in general. I think it is a general perception that old people are out of touch with modern times.

Other examples of prejudices were the perceptions that older people were waiting for their retirement and therefore not interested in hard work, were not committed to their jobs, they just sat there, nothing happened and they were difficult to co-operate with. A British woman said that "people tend to think that an older person in a job is quite happy just to toddle on and do that job which he has done for many years and maybe not develop things as much as a younger person would do".

Another disadvantage of older people concerned negative attitudes towards younger people. Several British women said that they sometimes found older people a bit reluctant to listen to younger people, to take them seriously, to give them credit, and they were not always very keen to accept younger people's ideas or waved their opinions aside. A British woman from the private sector said:

> Especially if you are a very young person and you want this progress and you think that you have this fantastic idea. It is quite demoralising when you can go to an older manager and say "We can do this and this" and he says: "Well, no, I do not think that would work."

A Swedish woman explained that there was a risk that older people might deprive younger people of their enthusiasm, "which is so dangerous to lose". Older people might also be negative to younger people's fast-moving careers. A British woman who had gone through her career path very quickly and now was at a senior level explained that "some of them, it is not all older colleagues, don't think it is right if you are young and you are senior and did not go through the path that they took".

Effects of prejudices against older people on younger people

Existing prejudices against older people seemed to have an effect on the lives of younger people. Some of the younger interviewees, especially those over 30 years of age, expressed their concern about being not very far off 40 and the risk of being overtaken by younger people when applying for a job and facing difficulties advancing due to age. One of the men said that for the age group 25-34 the age of 40 might not be that far away and "you want to be settled then". He continued:

> It is probably in the back of the minds of a lot of 25-35s that if
> it is my ambition to achieve something I need to be doing it
> now, because once I get to 40 it is too late. This is a crucial time
> in their careers.

Another man, aged 25, said that "the older you get the more opportunities have gone by and it concerns me that perhaps I am not making the most of the opportunities that I have got while I have got them".

Thus, these very well educated and ambitious younger interviewees seemed to feel pressured not to waste time and opportunities, to make the right choices and to move quickly in their careers because the time for a career was perceived to be short.

Younger and older managers

Several interviewees talked about the natural authority of older people and the natural respect that older managers inspired. Younger managers, on the other hand, could experience it as a problem when managing older people because of their lack of natural authority. It could be difficult to get older people to accept them as managers and to establish credibility and trust. This was especially the case when a person reporting to a young manager was "old enough to be my mother". It could result in a need to be very clear and calculated and to show who is the manager "especially if you are a young female manager, because people are going to see you as a secretary, not as a manager". None of the nine interviewees who had experience of having people much older than themselves reporting to them mentioned any advantages of managing older people. On the contrary, they found the situation to be more or less difficult and uncomfortable. A British woman concluded that a bad experience of managing older people could result in a gross generalisation, i.e., all older people were looked upon as a problem.

None of the interviewees had managers younger than themselves, but some of them had experience of having a manager of their own age. A Swedish woman said that she was more aware of power issues when the manager was young and she also felt that it was easier to be offended by a younger manager.

> When a younger manager asks me to do something for him, for
> example to go and pick up something that he could have picked
> up himself, I am more easily offended because he is young.

A 34-year-old British female manager reflected on what it would be like to have a younger manager. Her conclusion was that it would be challenging because she would interpret it as a personal deficiency and she would probably struggle, as it would signify "a fading of life beginning".

Intergenerational communication

Generally, the interviewees declared that they got along well with their older colleagues. Many interviewees, however, stressed the importance of common understanding, good social interaction, similar interests, exchange of ideas, the same lifestyle and understanding of life, and a similar sense of humour, all of which were more likely to occur between people of the same age. A Swedish man talked about younger and older people being in different life stages and therefore having different values and interests. This was especially obvious as long as the younger people had no children of their own. Another Swedish man expressed the following experience:

> I had nothing to discuss with the older people, except the job.
> Our interests were very different. They had grandchildren. It
> was boring. We had different values and we talked about things
> in different ways.

A British woman said that conflicts could get cleared away very quickly with younger people, "whereas with older people you do not know how to approach it". Generally, older people were perceived to be less approachable. A Swedish man expressed the following view:

> I feel that generally communication between generations is bad.
> People find it difficult to speak to people of other generations. I
> think it is true in working life, too.

Women and men

A British man from the private sector, aged 25, declared that he had more respect for older than younger female colleagues. Older female colleagues tended to be like another mother and look after the younger ones, whereas older men sometimes saw them as a threat. His conclusion was that older women were easier to work with. A British woman felt that older male colleagues tended to be a little paternalistic to her and more helpful. A Swedish woman held the view that older men might enjoy working with younger women and the prestige associated with them. On the other hand, young women might enjoy being fathered and like the sense of security older men give them. A British woman talked about the risks of parents at work. She thought it was very easy to fall into a parental and child role and that it worked both ways, but "it could create a dependency and not a relationship of equals".

A British man, who was a manager in a male-dominated industrial company, was convinced that there was a glass ceiling, i.e., an invisible barrier for women, in his company. He said:

> I think it is very difficult for women to get on in this organisation.
> I think older men, they are prejudiced against woman, so there
> is a lack of opportunities for women, because our managers are
> older.

Younger people's parents

Younger people's private experiences of older people might have an influence at work, for example on recruitment decisions. A Swedish woman said:

> When I think of conflicts I have had with my mother or with my
> parents-in-law, then I think I would not like to work with any of
> them. But I guess the conflicts are different as we are much closer
> to our members of the family than we are to colleagues and
> managers. But when in a position to employ, some people might
> be influenced by family conflicts.

On the other hand, parents may inspire respect for older people at work. A British man said that because his older colleagues were just slightly older than his father, he treated them with respect and how he would expect someone of his age to treat his father.

Practical measures for the encouragement of intergenerational co-operation

When asked about practical measures for the encouragement of intergenerational co-operation, a few tended to find no need for better intergenerational co-operation. However, when reflecting on the question, the interviewees presented many examples.

Several people suggested measures such as organising regular activities outside the workplace and getting people together, "so that they will have something in common to talk about". A Swedish man talked about the need for better communication between generations, as people spent most of their time with people of their own age. A British woman from a private company suggested that older people should be more interactive with the younger departments, as that could open up their minds.

Mentoring was another frequently suggested and appreciated measure. None of the organisations where the Swedish and British interviewees worked had any formal or organised mentoring programme, as far as the interviewees were aware of, and many of them regretted that. A few of the interviewees had previously had formal mentors. Generally, the mentor was supposed to be older and the protégé younger. A "reversed mentorship", i.e., a younger mentor and an older protégé was suggested by a British young man, whereas a Swedish woman came to the conclusion that to her a reversed mentorship would be "unnatural". A British male manager from a private company argued that the age gap between the mentor and the protégé must not be too large.

Team working in mixed teams and cross-age groups was considered important for many reasons, among others as a measure for the effective use of competence, for the prevention of stagnation, and last but not least for improving intergenerational relations at work. Lifelong learning and flexible career paths were other examples mentioned.

A Swedish woman talked about inadequate knowledge of the capacity of older people. She pointed out that there was a need to identify stereotypes and to help people give up their stereotypes by disseminating facts about older and younger people. Another Swedish woman suggested that employers should set down in writing in Human Resource Management policies the importance of age diversity.

DISCUSSION

The results presented above show that there exist stereotypes and prejudices against older people and that these may have an effect not only on older people but on people of other ages as well. Younger people's perceptions of what it is like to be older in working life may reinforce their behaviour concerning their own careers. They may feel the pressure not to waste time and opportunities and to move quickly in their careers before people younger than themselves overtake them. The fact that this stressful situation takes place at the same time as many younger people have young children indicates a possible negative effect on young children as well. In addition, as Fiske [11] points out, young people may be deprived of important models if they learn that their parents and grandparents are considered deficient and of little worth in working life.

However, stereotypes and prejudices against older people alone cannot explain why people risk being considered old in working life 20 years or more before retirement age. The results of the study suggest that intergenerational relations may be another clue to a better understanding of the phenomenon. Younger people may have negative experiences of older people's attitudes towards them or experiences of difficulties in having older people reporting to them. They may also have negative experiences of conflicts with other older people, for example parents. In addition, younger people sometimes know very few older people well, as they spend most of their time with people of their own age. Thus, various experiences of intergenerational relations or lack of experience may, together with stereotypes and prejudices, strengthen younger people's arguments for choosing younger applicants when in a position to employ.

These factors may also reinforce the youth culture and age segregation and act as a hindrance to a less ageist working life.

Furthermore, the results point to the importance of communication and good social interaction at work. The need for colleagues and managers to have a common understanding, speak the same way, and have the same interests must not be ignored when it comes to understanding intergenerational relations at work.

Mentoring, team working in cross-age groups, lifelong training, flexible career paths, and social activities outside the workplace were practical measures mentioned that may encourage intergenerational co-operation. Obstacles such as inadequate knowledge of the capacity of older people need to be cleared away and the advantages of age diversity need to be emphasised.

The above results did not show any significant differences between Sweden and the UK concerning intergenerational relations at work. The differences between individuals and workplaces seemed to be greater than the differences between the two countries.

CONCLUSION

A main conclusion is that ageism and age discrimination are phenomena that do not only concern a specific age group, i.e. older people. They may also have a decisive effect on the behaviour of younger people at work and on younger people's perceptions of their own career and future lives.

Another main conclusion is that a better knowledge and understanding of intergenerational relations at work would significantly add to the practical measures suggested above in order to improve intergenerational co-operation at work. An enhanced understanding of intergenerational relations could also contribute to a better use of people's skills and ability at work independent of age.

REFERENCES

1. Arrowsmith, J. and McGoldrick, A., 1996, Breaking the Barriers: A survey of managers' attitudes to age and employment, (London: The Institute of Management).
2. Itzin, C. and Phillipson, C., 1993, Age Barriers at Work: Maximising the Potential of Mature and Older People, (Solihull, UK: Metropolitan Authorities Recruitment Agency) .
3. Johansson, I., 1997, Ålder och arbete: Föreställningar om ålderns betydelse för medelålders tjänstemän, (Age and Work: Conceptions of the Significance of Age for Middle-Aged Employees), Doctoral Thesis, Department of Education, University of Stockholm.
4. Walker, A., 1997, Combating Age Barriers in Employment: European Research Report. (Luxembourg: Office for Official Publications of the European Communities).

5. Taylor, P. and Walker, A., 1996, Intergenerational relations in the labour market: the attitudes of employers and older workers. In *The New Generational Contract: Intergenerational relations, old age and welfare*, edited by Walker, A., (London: UCL Press), pp. 159-186.
6. Walker, A., (ed), 1996, The New Generational Contract: Intergenerational relations, old age and welfare, (London: UCL Press).
7. Glaser, B.G., 1978, Theoretical Sensitivity, (Mill Valley, CA: Sociology Press).
8. Coffey, A. and Atkinson, P., 1996, Making Sense of Qualitative Data, (London: Sage Publications).
9. Fontana, A. and Frey, J. H., 1998, Interviewing: the art of science. In *Collecting and Interpreting Qualitative Materials*, edited by Denzin, N. K. and Lincoln, Y. S., (London: Sage Publications), pp. 47-78.
10. HSFR (Humanistisk-samhällsvetenskapliga forskningsrådet. The Swedish Council for Research in the Humanities and Social Sciences), 1990, Forskningsetiska principer (Ethical recommendations), Stockholm.
11. Fiske, M., 1979, Middle Age: The Prime of Age? (Williemstad, Curacao: Multimedia Production).

16. A Work - Family Balance Approach to Research on Late Career Workers

Martin M. Greller[1], Linda K. Stroh[2]

[1] Management & Marketing Department, University of Wyoming, Laramie, WY, USA
[2] HRIR, Loyola University Chicago, Chicago, IL, USA

ABSTRACT

The work - family balance approach to inclusion in the work force is applied to late career workers. Implications for research and policy are discussed.

Keywords: *Late career, Work - family balance, Older worker*

INTRODUCTION

What would the work and aging literature look like if it were to take an approach similar to that adopted in the study of work - family balance? While work - family balance and work-and-aging have similar goals, their histories are quite different and, as a consequence, have taken two very different paths.

HISTORY OF THE TWO APPROACHES

Work - family balance research emerged from concerns about gender equity and an increased understanding of the effects of role conflict on women. While issues of balance certainly apply to men in the work force, as a women's issue the question was whether subtle forces were undermining the opportunities women had in the workplace [1] and if those forces were driving women to make choices based on influences other than their own inclinations [2]. Early work in this framework, identified a number of roles which competed for time (e.g., tasks at home) and individual identity (role as wife, mother, etc.) [3]. While there was no expectation that every women wished to invest herself wholly in career employment, a variety of factors were identified which would reduce the potential importance and personal investment in career.

For those women who did invest in establishing a career, the rewards were not always commiserate with the effort. Discrimination made it likely that women would not receive the same level of promotion, employment, and compensation as similarly situated male peers [4]. Thus, there were social barriers making it more difficult for a women to commit to a career, and the rewards for doing so were less.

While the areas for proposed interventions varied, their goal was relatively consistent: to increase options for women. The criterion for success would not be that more women worked in more demanding jobs, but that women were able to make the decision to pursue such opportunities with no greater barriers than might face similarly situated male counterparts and with the same expectation of rewards. With such ends in mind a number of interventions were tried. Some were social - cognitive, providing support either outside or within the employment setting, which could help interpret events and solve career related problems; others intervened into family relations. Another thrust was toward changing employers' practices - ranging from hiring, to training, to performance appraisal [5].

While the problems facing older workers may be similar (e.g., limited opportunity for employment and advancement, wage discrimination, social pressure to assume non-work roles), the approach to studying older workers and their employment has been quite different. Historically, that difference may be traced back to the perspectives of those who studied the employment issues of older workers.

Neo-classical economics has addressed the issue of employment in late career most directly. To some degree this was an effort to understand why a group of people who should have accrued the most human capital would be disproportionately out of the work force. The dominant explanation rests on three points which argue it is not worth training older adults [6]. Older workers cost too much to train - not that the training itself is more expensive, but their hourly wages are usually higher and (assuming they are more productive) the lost productivity will be greater. That greater cost cannot be recouped because older workers have less time left in their careers so there is not enough time to pay back the investment [7]. And, new skills acquired by older workers will be valued less in the market. This reasoning works as long as one accepts as axiomatic that workers over 50 will retire in the near future and that ageism (in wages offered older workers) is acceptable. If so, the logic works impeccably. If one does not accept these axioms then the explanation only serves to justify the refusal to invest in our most productive and valuable employees.

A second line of research was stimulated by social security. It is important for social security systems to have a reasonable estimate of the career behavior of older adults, particularly predicting how many of them will retire during a particular period. When older people cease to work they stop contributing to the social security system and eventually begin to draw from it. So, a number of studies have explored the antecedents of individuals' retirement decisions, and economic analyses explore the effect of different policies on aggregate retirement behavior [e.g., 8,9,10].

Industrial gerontology looks at the changing physical and psychological characteristics associated with aging and their implications for work [11]. While this may be done from a number of vantages and for a number of purposes, it is hard to escape the sense that the rate or degree of decline in older workers' performance, ability, or motivation is one object of the discussion. Even where the authors have a clear interest in extending careers, concern about the sustainability is a question.

Taken together these three traditions emphasize managing the exit of people, not developing their innate capacities. So, it should come as no surprise that the research has not focused on creating career continuity options for individuals.

While there have been repeated observations that the labor pool will be insufficient without the continued participation of its late career workers [12,13,14], this call has not been well heeded in the human resource management community. While there are certainly exceptions (e.g., Age Reform England), the traditional approach of organizations to their older employees has been to identify roles requiring reduced responsibility for the presumably diminishing motivation, capabilities, and human capital of older workers. Note the contrast to the way work - family balance tries to enhance the workforce participation opportunities.

WHAT COULD A WORK - FAMILY BALANCE APPROACH BRING TO THE STUDY OF LATE CAREER?

An approach similar to that used in work - family balance would add four dimensions to aging and work:

- Focus on the individual
- Support human capital development
- Identify barriers to opportunity and remove them
- A systemic view of workers and their environments

The question of focus should not be confused with one of discipline. Economics and sociology typically are associated with a macro view while psychology typically has a micro view. But, when it comes to aging and work, labor economics has broadened to include discussion of individual decision making. Conversely, psychological research has noted the social forces with which the individual must cope. So, an individual focus is not due entirely to discipline. More important is what question were being addressed.

An individual focus implies identifying paths by which individuals can realize their potential and aspirations. The important questions are: What will allow people in late career to maintain and use their human capital? Where social influences press people to define themselves as non-workers in late career, how can this pressure be offset so that an individual's own preferences may find more nearly free expression? It is the individual who makes the decision.

Contrast this to policy studies which ask by how much would the employment of 63 year olds increase if social security payments were reduced or delayed. There is little question that induced poverty can compel a proportion of older people to work. The question is the ratios. But, the criterion is an aggregate result. The objective of such research is to find ways to delay retirement. An individually focused approach would assume that there is a portion of the population that is inclined to continue employment if only the forces impeding them could be reduced. Those forces might be psycho-social, organizational, or based in social policy [15]. The assumption is that by removing enough barriers, the aggregate increase in employed 63 year olds would still be achieved. However, the human experience created by the intervention would be quite different. The specific 63 year olds who are at work may be different - as different as a group of conscripts compelled to work from a volunteer work force.

In seeking to support career continuity in late career, one of the major barriers is the lack of support for continued investment in human capital. This has become a problem because of the expectation of career ending and the intellectual justifications underlying it. Thus, we have a self-reinforcing process which appears to occur in late career whereby late career workers' skills need refurbishment. Yet, employers offer little encouragement and coworkers communicate subtle expectations that retirement will be the late career worker's next move. This results in late career workers participating in less training and development, which means they are less credible as candidates for advancement or new assignments. These diminished prospects lead them to see fewer opportunities in work, making retirement a more attractive option. Their departure from the work force validates the perceptions of the coworkers and employer, the very ones who originally discouraged the investment which would have kept the late career workers viable. It is a vicious circle [16].

A modest change in assumption could lead to profound change in outcome: Assume people are inclined to continue working unless they state otherwise. The decision belongs to the individual. Just as work - family balance makes no assumption about whether work or family is primary in an individual's concerns, the same could be done with late career workers with respect to work and retirement. It is not just a matter of asking what the individual's inclination is, but providing the individual with the information necessary to make an informed choice. This means including late career workers in the career counseling or career management process which occurs within the organization - rather than siphoning late career workers off to pre-retirement programs, a lesson we have learned from the work and family literature. This also means organizations will need to provide candid performance feedback to older workers. A similar logic might also be used to guide government sponsored programs for retraining.

Access to opportunity would seem to be assured by equal employment opportunity laws (in some countries) and the equity rights workers have to their jobs under the labor law (in other countries). However, when it comes to older workers the "rights" implied appear to be limited to holding onto one's current job. In practice, they may not extend to opportunities related to new duties, advancement, or development.

Social science is starting to entertain non-linear and recursive models (e.g., complexity theory, dynamic systems), and work - family balance has long tried to understand the interplay among the personal, interpersonal, organizational, and cultural factors [17]. Levinson [18] argued it was necessary to understand the full matrix of a person's life to appreciate how the various transitions were being addressed. While we do see discussion of "push " and "pull" factors when it comes to retirement decisions [19], there is a tendency to reduce the full richness of complexity and look at one variable at a time.

IS SUCH AN APPROACH APPROPRIATE?

It is certainly possible to argue for an approach to aging and work using a work - family balance approach, based on the similarity of the employment issues the two address. The employment issues are similar. But, is it anything more than an intellectual exercise to do so? The answer to this question rests on two points. Is it consistent with research on late career, and does it provide any potential benefit either to our understanding or ability to take action?

While advocates for older workers have noted the need to make better use of these people in the work force [12], in practice, people in late career received less training [20] and their employment prospects are more limited [21]. However, we are now beginning to see evidence that this is not the whole picture. People early in late career reply to surveys by saying they expect their work lives to continue well past traditional retirement age [22], and those who have left the work force express regret that they no longer have meaningful work to do [23]. So, at the level of sentiment there is reason to believe there is a pool of people in late career who have the inclination to work.

There also is reason to believe that social forces have discouraged older workers from continuing to maintain their skills and motivation. This is more subtle than the social aging process Hayward [24] criticized. Indeed, it is unlikely that coworkers tell a late career worker that they expect her to retire and then she does, despite a disinclination to do so. The process is apt to be more subtle, with the assumptions underlying the norms being built into the policies and procedures of our organizations, pension programs, and government policies; all of which make it easier to choose to exit the work force rather than to remain [15,25]. Work life continuity also appears to depend on the individual's perception of self-efficacy [26]. While self-efficacy is itself a psychological variable, it develops in response to the nature of the external environment (e.g., how challenging it is), social influence, actual personal ability and support, as well as self-esteem. This broad range of variables correspond to those which would be explored using a systemic framework.

There is also evidence, contrary to popular opinion, that older workers are participating in training [27]. This is usually interpreted as a change in late career behavior. However, we may have consistently underestimated the development undertaken by older workers because most of the previous research focused on employer-sponsored, on-the-job training [28]. Sterns [29] argued that older workers were most likely to participate in developmental experiences over which they exercised control. Simpson and her colleagues [16] found that late career workers do indeed participate in training, but primarily that which is under their own control and addresses specific career needs. When it comes to such training late career workers actually participate at higher levels than do younger workers. So, it is entirely possible that older workers had consistently invested in themselves at a far higher level than was acknowledged.

One of the questions which has been less well addressed is what sort of training or development is most beneficial to late career workers. However, we may be able to infer the answer from the factors that make older workers valuable to an organization. Where as training may enhance a younger worker's value by helping gaining mastery of the newest techniques or approaches, the value of the older worker comes from understanding how things work together. Tacit or context knowledge typically allows the older worker to anticipate problems, seeing implications not apparent to younger workers, and to solve problems, particularly those due to unintended consequences [30]. Indeed, one explanation for the observation that older workers seem to require more time to master training, is that they are doing more, hence the need to discuss and explore the implications of the training to a greater degree than younger workers [31]. This also suggests that the value delivered by older workers is highly individual, and an individual must be quite thoroughly invested in work to make these contributions.

In summary then, there are people interested in continuing their careers, they are at risk of being discouraged in ways which may be somewhat indirect, but if those impediments can be overcome (and, there is evidence that people do overcome them), then older workers may be active participants in the development necessary to keep them employable. They can do so by focusing on the activities necessary to enhance their unique contributions to their employers.

There are also points where work and aging differ from work - family balance. It is likely that most older workers will at some point choose to leave the work force; so while we can reduce barriers and provide support, eventually most will still choose to stop working. There seems to be little employer interest in fostering career continuity for older workers. Similarly, expanding employment opportunities for people in late career does not seem to be a political priority.

POLICY IMPLICATIONS

If we draw from work - family balance, we change the role of government from enactor to enabler. Policy becomes a tool for allowing individual action which might otherwise have been constrained. But, policy takes on life through the action of employers and the actions of supervisors and coworkers; so the actual manifestation of policy is the result of largely independent actions of groups and organizations. However, the independent actions are a source of innovation and best practices which can be observed and communicated broadly. In addition, independent organizations and groups can be sources of resistance to policy. The dynamics behind the resistance are also important to understand.

From a research perspective, this calls for studies which examine larger sets of interrelated variables (systems), cross level research (policy, organization, familial, and individual), which allows time for adaptation to occur and be observed. This is expressed pictorially in Figures 16.1 and 16.2.

What defines success? Currently, there are as many ways to define success in the area of aging and work as there are paradigms: better management of social security disbursements, increased numbers of older workers in the labor pool, more satisfaction in retirement, etc. While there is a temptation to say satisfaction with the balance of work versus other factors would be appropriate for aging as it has been in the work - family balance literature, a more appropriate, if more difficult, criterion may be actualization of human potential across the life span. After all, is that not what we would actually hope to foster?

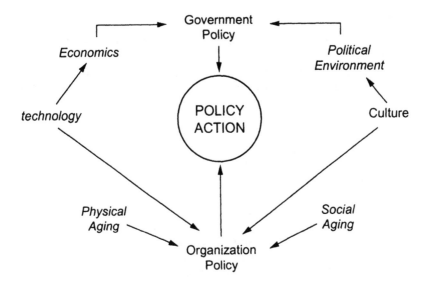

Figure 16.1 Traditional aging and work framework

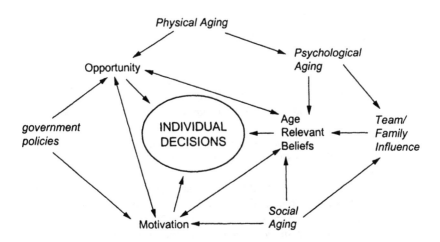

Figure 16.2 Systemic model for work and aging

REFERENCES

1. Mason, J.C., 1993, Knocking on the glass ceiling. In *Management Review*, **82** (July), pp. 5.
2. Stroh, L.K., Brett, J.M. and Reilly, A.H., 1992, All the right stuff: A comparison of female and male career patterns. In *Journal of Applied Psychology*, **77**, pp. 251-260.
3. Yogev, S. and Brett, J.M., 1985, Patterns of work and family involvement among single and dual earner couples: Two competing analytic approaches. In *Journal of Applied Psychology*, **70**, pp. 754-758.
4. Powell, G.N., 2000, The Glass Ceiling: Explaining the good and bad news. In *Women in management*, edited by Davidson, M.J. and Burke, R.J., (London: Sage Publications), pp. 236-249.
5. Nelson, D.L. and Burke, R.J., 2000, Women executives. In *Health, stress, and success Academy of Management Executive*, **14**, pp. 107-121.
6. Mincer, J., 1994, Human capital: A review. In *Labor Economics and Industrial Relations: Markets and Institutions*, edited by Kerr, C. and Standohar, P.D., (Cambridge, MA: Harvard University Press), pp. 109-141.
7. Rix, S.E., 1996, Investing in the future: What role for older worker training? In *Handbook on Employment and the Elderly*, edited by Crown, W.H., (Westport, CT: Greenwood Press), pp. 304-323.
8. Gendell, M. and Siegel, J.S., 1996, Trends in retirement age in the United States, 1955-1993 by sex and race. In *Journal of Gerontology: Social Science*, **51B**, S32-S39.
9. Gustman, A.L. and Steinmeier, T.L., 1985, The 1983 social security reforms and labor supply adjustments of older individuals in the long run. In *Journal of Labor Economics*, **3**, pp. 237-253.
10. Rust, J. and Phelan, C., 1997, How Social Security and Medicare effect retirement behavior in a world of incomplete markets. In *Econometrica*, **65**, pp. 781-821.
11. Sterns, H.L. and Miklos, S.M., 1995, The aging worker in a changing environment: Organizational and individual issues. In *Journal of Vocational Behavior*, **47**, 248-268.
12. Brewington, J.O. and Nassar-McMillan, S., 2000, Older adults: Work-related issues and implications for counseling. In *Career Development Quarterly*, **49**, pp. 2-15.
13. Greller, M.M. and Nee, D.M., 1989, From baby boom to baby bust: How business can meet the demographic challenge. In *Reading*, (MA: Addison-Wesley).
14. Johnston, J.W. and Packer, A., 1987, Workforce 2000: Work and workers for the Twenty-First Century. In *Indianapolis*, (IN: Hudson Institute).
15. Greller, M.M. and Stroh, L.K., 1995, Careers from mid-life and beyond: A fallow field in need of sustenance. In *Journal of Vocational Behavior*, **47**, pp. 232-247.
16. Simpson, P., Greller, M.M., and Stroh, L.K., (in press), Variation in human capital investment activity by age. In *Journal of Vocational Behavior*.
17. Kossek, E.E., and Ozeki, C., 1998, Work-family conflict, policies, and job-life satisfaction relationships: A review and directions for organizational behavior-human resources research. In *Journal of Applied Psychology*, **83**, pp. 139-149.

18. Levinson, D.J., 1986, A Conception of Adult Development. In *American Psychologist*, **41**, pp. 3-13.
19. Schultz, K.S., Morton, D.R. and Weckerle, J.R., 1998, The influence of push and pull factors on voluntary and involuntary early retirees' retirement decision and adjustment. In *Journal of Vocational Behavior*, **53**, pp. 45-57.
20. Frazis, H., Gittleman, M. and Joyce, M., 1998, Determinants of training. In *An analysis using both employer and employee characteristics*, (Washington, DC: Bureau of Labor Statistics).
21. Chai, S. and Stevens, A.H., 2001, Job loss and employment patterns of older workers. In *Journal of Labor Economics*, **19**. pp. 484-521.
22. American Association of Retired Persons, 1989, Baby boomers envision their retirement. In *An AARP segmentation analysis*, (Washington, DC: Author).
23. McNaughton, W., Barth, M.C. and Henderson, P.H., 1989, The human resource potential of Americans over 50. In *Human Resource Management*, **28**, pp. 455-473.
24. Hayward, M.D., 1998, Commentary: The usefulness of age norms in retirement research. In *Impact of work on older adults*, edited by Schiae, K.W. and Schooler, C., (New York: Springer), pp. 124-130.
25. Greller, M.M., 1999, Age norms and career motivation. In *International Journal of Aging and Human Development*, **50**, pp. 213-224.
26. Mauer, T.J., 2001, Career related learning and development, worker age, and beliefs about self-efficacy for development. In *Journal of Management*, **27**, pp. 123-140.
27. Tuckett, A. and Sargent. N., 1996, Headline findings on lifelong learning from the NIACE/Gallup Survey 1996. In *Adults Learning*, **7**, pp. 219-223.
28. Organization for Economic Cooperation and Development, 1990, In *Employment outlook*, (Paris: Author).
29. Sterns, H.L., 1986, Training and re-training adults and older adult workers. In *Age, health, and employment*, edited by Berren, J.E., Robinson, P.K. and Livingston, J.E., (Englewood Cliffs, NJ: Prentice-Hall), pp. 93-113.
30. Greller, M.M. and Simpson, P., 1999, In search of late career: A review of contemporary social science research applicable to the understanding of late career. In *Human Resource Management Review*, **9**, pp. 103-120.
31. Sterns, H.L. and Doverspike, D., 1988, Training and developing the older worker: Implications for human resource management. In *Fourteen steps in managing an aging work force*, edited by Dennis, H., (Lexington, MA: D.C. Heath), pp. 97-110.

17. Work Climate and the Age-Hostile Workplace

Jacqueline Agnew[1], Gilbert C. Gee[2], David J. Laflamme[1],
Karen A. McDonnell[1], Barbara A. Curbow[1]

[1] *Johns Hopkins Bloomberg School of Public Health, Baltimore,
Maryland, United States*
[2] *Indiana University-Bloomington, Indiana, United States*

ABSTRACT

Telecommunications work is characterized by strict supervisory controls and highly structured policies. This study examined the work climate characteristics of telecommunications worksites in which policies are perceived to be differentially enforced according to worker age. The sample consisted of 276 U.S. telecommunications organizations primarily related to customer service. Surveys addressed single work locations and addressed workforce description, policies, policy enforcement, work conditions, and reporter characteristics. Almost 18% of the workplaces were classified as "age-hostile," i.e., age of over 55 years was perceived to "hurt the worker's case" in a policy violation. Compared to age-neutral/friendly workplaces, these tended to have greater proportions of workers under the age of 40. Age of reporter was not associated with perceptions of workplace hostility. Almost all measures of work climate (management actions, work conditions, communication, and supervisor relationships) were less favorable for age-hostile work locations, with the most striking differences seen for supervisor relationships. Worker morale, job satisfaction, and job security were also strongly negatively associated with age-hostility. Policies and the manner in which they are enforced through managerial and supervisory actions reflect the workplace climate of a work site. This indirect form of age discrimination may be influenced by the age structure of a workplace and merits further research attention.

Keywords: *Aging workers, Telecommunications, Work climate, Workplace policies*

INTRODUCTION

Although discrimination takes many forms, age discrimination has been commonly noted in the workplace and was found to be the most common form of discrimination in the European Union.[1] Age discrimination may be practiced overtly or more subtly through personnel actions such as hiring, firing, training, or promotion, and it is thought that most age discrimination is indirect. While workplace policies generally are put into place to ensure standardized treatment of workers, the actual outcomes for those who violate those policies may not be consistent.

The telecommunications industry is one example of a work environment characterized by rigidly structured policies established by contracts between unions and management. Telecommunications workers are subject to frequent monitoring for surveillance over their behavior, including listening to their conversations with customers, directly observing them at work, or monitoring behavior electronically by timing and counting the number of calls handled. If a policy is violated, the consequences can be severe, sometimes leading to dismissal or suspension without pay.

This study of telecommunications workplaces in the United States examined whether policies were perceived to be enforced differentially according to worker age. Additionally, factors associated with such "age-hostile" policy enforcement were identified. The study was carried out as part of a three phase investigation to look at associations between policies, substance use, and worker well being. The research questions were: 1) Is there differential enforcement of policies according to worker age? and 2) What are the characteristics of an "age-hostile" environment?

METHODS

Participating workplaces were entered into the study by contacting the presidents of each union local where the membership performed primarily customer service jobs. Each president was asked to nominate a respondent for that local - one who was most knowledgeable about the workplace, workers and policy. We recommended that the chief union steward be considered first, but others could be considered at the president's discretion. A survey questionnaire and return envelope were mailed to each nominee. Instructions indicated that the respondent should consult any useful accurate sources of data to answer the survey questions. If more than one worksite was represented by the local, the nominee was asked to report on only one - the site with which he or she was most familiar. A total of 535 surveys were mailed, and 276 responded, for a response rate of 52%.

The questionnaire had been developed with the assistance of a formal advisory board to the project made up of representatives of workers, labor officials, and management from the telecommunication industry as well as researchers who were not part of the study team. It included questions on demographics, policies, perceived work conditions, respondent characteristics, and perceived bias with regard to policy violation outcomes.

Respondents were asked to report the demographic characteristics of their respective workforces at their sites. Based on their responses, over half (57.8%) of the workers represented were over 40 years old. Only about one fourth (24.1%) were non-white. The ratios of women to men were about equal (50.6% women), and the average estimated size of a workplace was 86 workers (SD 35.7). The respondents themselves very nearly reflected this picture, with an average age of 47.4 (SD 6.3), equal numbers of males and females, but a lower percentage of non-whites at (11.1%). They had many years of experience - 24.9 (SD 7.1) on average.

A workplace was considered to be "age-hostile" if the respondent felt that being over age 55 would lead to a negative outcome if that older person was thought to violate a workplace policy. The questionnaire asked: "In the event of a worker being charged with violating management policy due to observation by a supervisor or electronic or computer monitoring, is there anything about that worker that might influence what happens?" Respondents were asked to select whether being over age 55: helps the worker's case; hurts the worker's case; or neither helps nor hurts. In another section of the survey, the question was asked: "In the event of a worker failing an alcohol or drug test, is there anything about that worker that might influence what happens?" Response choices were the same. It should be noted that older age was included among several other possible types of policy enforcement biases such as ethnicity, seniority, and gender.

The definition of "age-hostile" was based on the combined responses to the two questions about treatment of workers accused of performance monitoring and substance use policy violations. Age over 55 years was considered to hurt one's case for either type of violation at 46 worksites or 17.7% of the sample. These sites were compared to the remainder of the sample, the age neutral or age friendly worksites, on the following variables as well as overall age structure of the workforces.

Perceived work conditions were evaluated by three sets of questions directed at overall climate as well as treatment of workers by management and supervisors. Perceptions of worker morale, job satisfaction, job security, and job stress were measured by asking the respondent to select a rating on a scale of 1-10 to indicate low to high levels. Characteristics of management and perceptions of management were explored with questions on communications, commitment to quality, interest in worker welfare, and fair treatment of workers. Responses were based on a five point Likert scale of frequencies ranging from never true to always true. Five additional questions addressed supervisor actions. Again, responses were on a five point Likert scale ranging from never to always.

RESULTS

Figure 17.1 shows results on questions regarding policy outcomes. The upper box refers to policy violations based on performance monitoring. While 12.7% said being over age 55 helps one's case, 16.3% indicated this would hurt one's case. Seventy-seven percent said older age would neither help nor hurt. When asked about policy violations for drug and alcohol, the pattern was reversed. While 13.6% said older age would help one's case, only 11.5% said it would hurt one's case. Seventy-five percent said older age would neither help nor hurt. This reversal may be because drug and alcohol use is often seen as a problem of the younger employee, and the older worker might be considered more credible in contested cases.

Findings for morale, job satisfaction, and job security indicated that age friendly/ neutral sites were rated significantly higher in each area, as shown in Figure 17.2. While job stress levels were also felt to be higher in the age friendly environments, this difference was not significant. Ratings for stress, however, are quite high - exceeding 8 for both groups. The greatest difference between both groups was seen for job security.

Being over age 55 ...

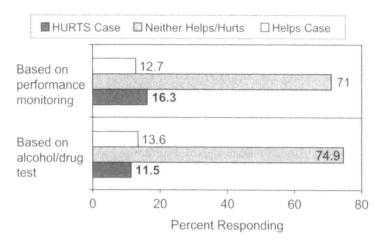

Figure 17.1 Outcomes for violating policy

Figure 17.2 Worker perceptions

Figure 17.3 shows that all differences between group ratings on management characteristics were significant, with age-hostile workplaces scoring lower. The largest difference was seen in response to management's interest in the welfare of its people. Similarly, Figure 17.4 indicates consistently significant differences between the groups on perceptions of the work climate established by supervisors.

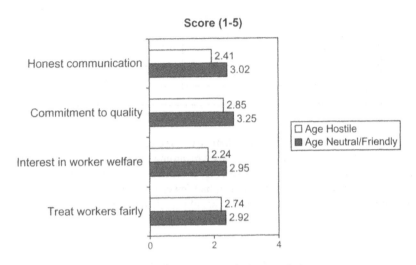

Figure 17.3 Management characteristics

Figure 17.4 Supervisor actions

The age structure of the worksites did differ by group, as shown in Table 17.1, with larger proportions of workers aged 26-40 in the age-hostile group. In the age friendly worksites, a larger proportion fell between ages 41 and 55, and a slightly smaller proportion were older than 56.

Table 17.1 Age composition of workplaces

	Age Hostile	Age Neutral/Friendly
<26 years	15.5	14.2
26-40	35.8	27.7
41-55	37.2	49.9
56+	11.5	8.3

Finally, we asked the question whether the age or other characteristics of the respondent might have influenced responses regarding age bias in policy enforcement outcome. Findings, as seen in Table 17.2, indicated that respondents for the age hostile workplaces did not differ in age, had fewer years in the industry, and were more typically female. Differences in years of experience may have been associated with gender as it is more common for women to begin their careers at an older age due to family obligations.

Table 17.2 Respondent characteristics

	Age Hostile	Age Neutral/Friendly
Age (mean)	46.8	47.1
Years in Industry	22.7	25.3 **
Gender (% female)	63%	48% *

$**p=.03$ $*p=.07$

CONCLUSIONS

These results suggest that, in the rigidly structured environment of the telecommunications industry, differential policy enforcement according to age constitutes a possible form of age discrimination. While policies generally are established to ensure clear expectations and standardized treatment for violators, subgroups of a population may suffer if enforcement varies according to specific characteristics. This type of discrimination, especially when directed at older age groups, is difficult to objectively measure and document. Consequently, it may be quite difficult to prevent.

For this sample of worksites, differential policy enforcement for older workers was also associated with indices of work climate and worker well being. Of note were the findings that management and supervisors at age-hostile sites were seen as less concerned about worker well being and less likely to communicate well with workers. Others have pointed to the importance of the design and management of work and work organization and its potential role in work-related stress, particularly among older workers.[2] More information is needed about other characteristics of organizations and work environments, the behavior of supervisors and managers, and worker well being in order to better understand the causal relationships involved.

In age-hostile workplaces a greater proportion of workers are under age 40 suggesting that younger workers may contribute to an atmosphere of age hostility. This may be related to an overall climate within an organization based on prevailing attitudes of the majority, or again may driven by behaviors or beliefs of younger managers and supervisors. Despite evidence to the contrary, surveys have demonstrated that employers continue to believe that the work performance of older workers is not as acceptable as that of their younger counterparts.[3]

More work is needed to look at policy enforcement differences as a potential form of age discrimination. It will be especially important to further examine the suggestion that treatment of older workers is related to various work climate measures and to consider interventions that might better ensure a fair workplace for all.

Acknowledgements: This work was supported by Grant Number 034902 from the Robert Wood Johnson Foundation, Substance Abuse Policy Research Program. We would like to acknowledge the support of the union, Communication Workers of America, as well as our additional team members, Joan Griffith, David LeGrande, and Margaret Ensminger.

REFERENCES

1. Ilmarinen, J., 1999, Ageing Workers in the European Union-Status and Promotion of Work Ability. In Employability and Employement, Helsinki, Finland, edited by Painotalo Miktor Ky, pp. 233.
2. Griffiths, A., 2000, Designing and managing healthy work for older workers. In *Occupational Medicine*, **50**(7), pp. 473-477.
3. Wegman, DH., 1999, Older workers. In *Occupational Medicine, State of the Art Reviews* **14**(3), pp. 537-557.

18. Occupational Activity and Aging

Sheng Wang

*Department of Occupational and Environmental Health,
Peking University Health Science Center, Beijing, China*

ABSTRACT

Studies conducted in China in the field of work and aging were reviewed. If workers are exposed to chemicals or other occupational hazards over a long period, their work ability may be adversely affected and such exposure may cause aging. It is still necessary to explore evaluation methods for aging.

Keywords: *Work, Aging, Occupational hazard*

Occupational activity is important to human life and to social development. In modernized production, people may be exposed to physical agents and/or chemicals during their work. The effects of stress induced by high frequency repetitive tasks on psychology and physiology are becoming a significant occupational issue. Scientists all over the world are paying more attention to the relationship work and aging. This study reviews research by Chinese in the area of work and aging.

Zhou Tong [1] investigated 187 workers and 161 controls by using the work ability index (WAI). The empolyees worked in a paint factory and the concentration of benzol and toluene at the workplace were 36 mg/m^3 and 90 mg/m^3 respectively. Results showed that the WAI of workers was lower than that in control groups, especially in workers aged over 45 years. (Figure 18.1, Figure 18.2). Statisticals analysis of the relationship between WAI and work age showed that WAI decreased quickly in workers (Table 18.1).

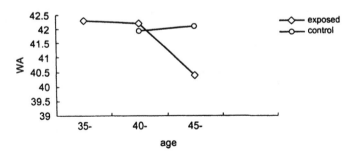

Figure 18.1 Comparison of WAI in male

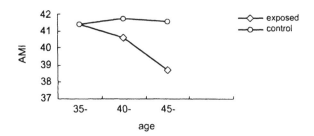

Figure 18.2 Comparison of AMI in female

Table 18.1 The correlation of WAI with age and work age

	WAI of exposed						WAI of control					
	Male (n=63)		Female (n=115)		Total (n=178)		Male (n=46)		Female (n=108)		Total (n=154)	
	(r)	(b)	(r)	(b)	(r)	(b)	(r)	(b)	(r)	(b)	(r)	(b)
Age (a)	-0.14	-0.142	-0.22*	-0.298	-0.16*	-0.204	-0.15	-0.097	-0.21*	-0.167	-0.15	-0.08
Work age(a)	-0.22	-0.326	-0.18	-0.102	-0.12	-0.067	-0.09	-0.035	-0.31**	-0.150	-0.22**	-0.09

r: correlation coefficient; b: gradient; * $P<0.05$, ** $P<0.01$ vs control

In order to study the effects of benzol solvents on aging, experimental study of toluene on the life span of the fruitfly was done in the laboratory. Orepon/C strain fruitflies (male and female) were cultured in tubes with toluene at concentrations of 0 mg/m³, 100 mg/m³, 375 mg/m³, and 1000 mg/m³ until all fruitflies died. The results showed that the life span, including average, longest and shortest life span were shortened in exposed groups of both sexes and there was significant difference. The average life span decreased 12.9%, 15.4%, 22.7% in males and 7.5%, 13.7%, 26.6% in females (Table 18.2). It suggested that toluene exposure accelerated the aging of fruitflies and shortened their life span [2].

Table 18.2 The effect of toluene on life span(day) of fruitfly

sex	Concentration (mg/m³)	n	Average life span	Longest life span	Shortest life span	Median life span	Average longest life span	Average shortest life span
female	0	160	64.4 ± 12.7	77	23	68	77.0	35.2
	100	160	56.1 ± 13.1**	77	18	57	75.2	28.3
	375	160	54.5 ± 12.8**	72	16	58	70.8	26.0
	1000	160	49.8 ± 13.4**	68	13	52	66.5	20.6
male	0	160	59.8 ± 12.7	79	23	60	77.8	32.2
	100	160	55.3 ± 15.5**	77	11	60	76.4	17.5
	375	160	51.6 ± 14.4**	75	9	55	73.2	19.5
	1000	160	43.9 ± 12.9**	72	9	44	48.8	13.2

VS control group ** $P<0.01$

103 Al-exposed workers and 64 controls aged 25-60 years were selected for investigation of Al-induced neurotoxic effects and age dependent differences in neurobehavioral function by using the WHO-neurobehavioral core test battery (NCTB) [3]. The air concentration of aluminium in workplace was 5.31 ± 2.85 mg/m^3. The content of inhaled aluminium per work day was 28.68 ± 21.62 mg/m^3 and urinary aluminium of workers was 30.36 ± 18.86 ug/L. The results showed that the elderly Al-exposed workers had obvious mood swings, i.e., scores for tension-anxiety, depression-dejection and anger hostility were significantly higher than those of the control group of the same age ($P<0.05$). Digit span performance in young Al-exposed workers, digit symbol score in Al-exposed workers of middle age, and pursuit aiming dot of elderly Al-exposed workers were markedly different from those of controls of corresponding ages. It was concluded that occupational aluminium exposure might interfere with normal neurobehavioral function and the effects seemed to be age-dependent. Young workers showed memory loss. The elderly workers showed mood disorder and decline of motor speed and accuracy (Table 18.3, Table 18.4).

Ma Laiji studied physiological age (PhA) and psychological age (PsA) of 179 industrial workers, 184 teachers and 174 enterprise administrators by using the human aging instrument (HAI). The results indicated that the percentages of individuals with older agePhA than calendar age (CA) in industrial workers, teachers and administrators were 50.84%, 27.72% and 12.26%, respectively. The percentage with older PsA than CA in the three groups were 68.16%, 45.11% and 44.25% respectively. There were significant differences in percentages of older PhA or PsA than CA among the three groups. There was a tendency toward increasing PhA and PsA in industrial workers as compared to teachers and enterprise administrators [4] (Table 18.5).

Table 18.3 Results of neurobehavioral tests in Al-exposed and control groups (X±S)

Item	Exposed			Control		
	25~34a (n=49)	35~44a (n=33)	45~60a (n=21)	25~34a (n=23)	35~44a (n=22)	45~60a (n=19)
simple reaction time (ms)						
Mean	288.1±76.2	333.6±106.7	367.9±96.0	267.2±55.4	298.5±59.0	341.6±107.5
Digit span score						
Forward	11.6±0.2*	11.3±2.6	11.0±2.4	12.6±1.3	11.5 ±2.3	11.0±2.1
Backward	5.8±3.0	4.5±1.6	4.2±1.7	5.7±2.6	5.5±3.0	4.5±1.5
Santa Ana score						
Preferred hand	39.0±4.7	36.8±5.8	32.8±5.6	39.9±6.2	39.9±6.0	35.7±5.6
Non-preferred hand	37.7±5.1	34.2±5.5	31.0±5.0	37.0±5.6	35.8±7.3	33.3±4.8
Digit symbol score	51.9±12.3	37.1±7.8**	33.8±7.8	52.1±9.7	46.3±10.6	36.2±12.4
Benton visual retention score	7.1±1.9	6.6±2.1	5.9±1.7	7.7±1.4	7.5±1.6	6.7±1.5
Pursuit aiming dot						
Total	212.5±39.9	187.2±27.1*	171.7±7.9*	222.0±9.4	209.9±38.4	200.2±47.5
Correct	188.8±44.8	175.6±23.3	155.3±29.1	208.9±7.8	191.4±39.5	173.6±42.4
Error	19.3±31.5	11.1±10.3	14.8±15	17.6±5.5	13.7±13.4	26.7±18.5

Adjusted by year of education and employment, expressed as adjusted means ± deviation; * $P<0.05$, ** $P<0.0$ compared with control of corresponding age

Table 18.4 Survey of profile of mood state (POMS) in Al-exposed and control groups (score, X±S)

group	Tension-anxiety			Depress-dejection		
	25~34a	35~44a	45~60a	25~34a	35~44a	45~60a
exposed	11.7±7.1(49)	14.5±6.9 (33)	15.2±10.6(21)*	16.9±11.4(49)	20.0±12.9(33)	23.0±16.6(21)*
control	8.8±4.7(23)	13.7±6.8(22)	8.6±6.0(19)	13.0±7.6(23)	18.2±14.8(22)	12.6±9.4(19)

group	Anger-hostility			Vigor-activity		
	25~34a	35~44a	45~60a	25~34a	35~44a	45~60a
exposed	15.1±9.3(49)	17.4±9.8(33)	20.5±12.9(21)*	20.3±6.8(49)	20.1±5.4(33)	19.7±7.0(21)
Control	11.3±6.2(23)	17.2±10.8(22)	11.6±6.5(19)	21.6±6.3(23)	20.5±6.5(22)	19.7±5.5(19)

group	Tension-anxiety			Depress-dejection		
	25~34a	35~44a	45~60a	25~34a	35~44a	45~60a
exposed	9.3±5.6(49)	12.5±5.0(33)*	13.0±6.4(21)**	9.9±5.3(49)	12.0±4.8(33)	12.0±5.3(21)
control	6.7±5.0(23)	9.0±6.4(22)	7.1±4.9(19)	8.0±3.6(23)	9.6±5.7(22)	10.1±4.0(19)

Adjusted by year of education and employment, expressed as adjusted means ± standard deviation; * P<0.05, ** P<0.01 compared with control of corresponding age; figures in the parentheses are the number of subjects

Table 18.5 The proportion of older PhA and PsA compared to CA in occupational classifications

Different group	n	PhA>CA		PsA>CA	
		n	%	n	%
Industrial workers	179	91	50.84**	122	68.16**
Teachers	184	51	27.72**	83	45.11
Enterprise administrators	174	22	12.64	77	44.25

PhA: physiological age, PsA: psychological age, CA: calendar age. X^2 test ** P<0.01)

Evaluation methods and index of work ability and health status for aged people are important and require further investigation. Experimental and field studies were conducted using testing life capacity index (LCI), physical index (PhI), and mental function including short memory, decode and delete letter. Results demonstrated that PhI was steady and simple. The indices, including LCI, PhI and mental function may be beneficial for assessing health status of aged people objectively [5]. Feng's research combined physical capacity and mental function as an aging index. After it was used to analyze 559 workers in different age groups, the result showed that the aging index could make exact distinctions for different aged people [6].

Zhu Zhiming et al investigated 555 aged workers using questionnaire promulgated by the Chinese Geriatric Association, and the data were treated statistically using the stepwise regression method. The results showed that senior citizens with low quality of life scores obviously were influenced by physical health status and income, which affect each other. The psychological well-being and living habits were ignored [7].

REFERENCES

1. Zhou, T., Jin, X. and Jia, X. *et al.*, 1998, Aging effects of low level benzol solvents on working population. In *Journal of Labour Medicine*, **15**(3), pp. 129-133.
2. Wang, H., Zhou, T. and Ma, L. *et al.*, 1998, The effect of toluene at low concentration on life span of fruitfly. In *Journal of Labour Medicine*, **15**(3), pp. 134-136.
3. Guo, G., Ma, H. and Wang, X. *et al.*, 1999, Age-dependent difference of neurobehavioral function among workers exposed to aluminium. In *Chinese Journal of Industrial Hygiene and Occupational Disease*, **17**(2), pp. 74-76.
4. Ma, L., Jin, X. and Zhou, T. *et al.*, 2000, *Effects of occupational classifications on physiological and psychological age*, **18**(3), pp. 155-157.
5. Wang, M., Zhan, X. and Zhan, C. *et al.*, 1997, Study on health assessmental index and method of aged people. In *Journal of Occupational Health and Damage*, **12**(2), pp. 68-71.
6. Feng, Q., Fang, Y. and Li, X. *et al.*, 1995, Determination and measurement of the degree of human aging. In *Chinese Journal of Gerontology*, **15**(4), pp. 211-213.
7. Zhu, Z., Zhou, Y. and Zhong, S. *et al.*, 1997, Quality of life of elderly workers in Changsha Urban district. In *Chinese Journal of Gerontology*, **17**(5), pp. 262-264.

PART IV
Maintaining Work Ability of Elderly Workers

19. Company-Level Strategies for Promotion of Well-Being, Work Ability and Total Productivity

Ove Näsman

Occupational Health Service, Fundia Wire Oy Ab, Dalsbruk, Finland

ABSTRACT

In many companies in Finland, employers and employees have reached a consensus: well-being, work ability and total productivity are closely linked together. The knowledge on "Aging and Work" is vast enough today for companies to start interventions for promotion of well-being and work ability among aging workers, aiming also at improved total productivity for the company.

To achieve the best results at the company level collaboration between skilled researchers and enthusiastic employees is needed. Working in this way, the company can have access to good scientific knowledge and also has the certainty that the process will continue among the employees after the research project is finished.

The Occupational Health Service (OHS) at Fundia Wire has carried out many projects in collaboration with different partners, but mainly the Finnish Institute of Occupational Health (FIOH).

The most important projects have been DalBo, Respect for the Aging and Metal-Age.

The projects have had a huge impact on well-being, work ability and total productivity.

Experiences from these projects are included in Fundia Wireys, which is a comprehensive program for promoting work ability.

Keywords: *Well-being, Work ability, Total productivity*

INTRODUCTION

Fundia is a Nordic steel company and a member of the Finnish Rautaruukki Group. Fundia produces long steel products such as wire and bars. Fundia Wire Oy Ab in Finland has one blast furnace and one rolling mill with 600 employees having a mean age of 44.2 years. The annual monetary investment in the OHS is about 500 Euro/employee.

At Fundia, the OHS has a long tradition of close collaboration with both management and employees. During the last decade there has also been intensive collaboration with external partners, mainly researchers from FIOH (1-2).

THE FUNDIA WIREYS PROGRAM

Fundia Wire's program for promotion of work ability is called Fundia Wireys (Figure 19.1). Wireys is Finnish for "alertness."

Figure 19.1 The Fundia Wireys program for promotion of work ability

The Fundia Wireys program is a process continuing throughout the working life, from employment to retirement. In our view, the best results concerning work ability and well-being of aging workers are achieved when all employees, both young and old, participate actively. Well-being in old age is usually established in the younger years, and therefore activities should start already among relatively young employees. It is important to view the promotion of work ability as a whole, and therefore all the different projects and campaigns are part of this Fundia Wireys program. The OHS is responsible for implementing most aspects of Fundia Wireys.

In Fundia Wireys, on the right hand side of the arrow we have activities based on age or years of service, starting from the bottom with a thorough employment health check-up. This check-up is always done before the employment contract is signed.

The basic course on maintenance of work ability is a 12-day course at a rehabilitation centre. This course is similar to the earlier DalBo course at Fundia during 1990-1994, which was very popular among the employees and also proved to be a success economically. The business economics analysis of the DalBo project showed that the annual benefits were ten times the annual costs(3).

The aim of the basic course is for the participants to know, upon completion of the course, the fundamentals of individual promotion of work ability and well-being. Important parts of the course are a variety of physical activities, ergonomics including analysis of each participant's work video, health check-up including muscle tests and walking test, psychology, nutrition, care of the feet and relaxation. The follow-up after the course will be five years with annual tests except the first year, when participants are tested twice. In the future every employee at Fundia Wire will have the right to participate in this course once during his/her active working life. If the employee is young at the time of employment and has no health problems, it may be best to attend the course after, for example, 15 years of employment. If the employee is older at the time of employment and already has some health problems, he/she can participate in the course after only three years. The first course will take place in the autumn of 2001.

The course "Redirecting resources" is planned for aging employees, who still have 10-15 years of work before retirement. At this point in one's working life, something often is required to keep up one's motivation to work. This course will be planned individually and could be, for example, 2+2+2 days with six months between the 2-day periods. This course activity will start at the end of 2001 or the beginning of 2002.

Every employee has a regular health check-up every fifth year up to 50 years of age, and after that every third year.

Every aging employee at Fundia has the opportunity to join the 5-day Sundia course at a rehabilitation centre three times. This happens during the years after the employee's 54th, 59th and 63rd birthdays. The program at the Sundia course consists of physical activities, relaxation, massage, psychology and discussions. The Sundia course started in 1995 as part of the Respect for the Aging program. This program had four sub-projects:

TA: Changes at work and the adaptation of older workers
Fun-shift: Developing a three-shift system
Kelaus: Reducing the physical handling of heavy materials
Åminnefors: Developing collaboration between employees of different ages

The activities on the left side of the arrow in Fundia Wireys continue throughout the entire working life irrespective of age or years of service.

Basic tasks of the OHS:

The OHS at Fundia (physician, nurse, and physiotherapist) provides medical treatment at general practitioner level. By treating all kind of medical problems, the OHS personnel get to know all the employees and also create a relationship of trust with most of the employees. Care of the sick occupies approximately one-third of the working time of the OHS personnel.

Ergonomics is an important part of Fundia Wireys and is mainly handled by the physiotherapist, who is also responsible for activities concerning physical exercise and pause gymnastics.

The Intoxicant programs have been implemented actively in recent years. The attitude towards alcohol abuse is changing among employees, and the former silence is now changing into active talk on helping coworkers who have alcohol problems. Smokers can obtain nicotine substitutes from the OHS when they try to stop smoking. The COPD project is a campaign for heavy smokers with high risk of COPD (Chronic Obstructive Pulmonary Disease).

Work place investigations are statutory in Finland and comprise a comprehensive analysis of different aspects such as physical and mental load, exposure to harmful agents, etc., and suggestions for improvements.

The statutory health controls at Fundia are mainly audiograms because of the high level of noise, and chest x-rays because of exposure to asbestos in earlier decades.

The DalBo follow-up weekend every third year is planned as a leisure time activity, involving mainly physical activities at a sports or rehabilitation centre. The employer pays the costs. Before one can participate in this weekend, participation in either the former DalBo course or the Basic course on maintenance of work ability is required. The first weekend will take place in autumn 2001.

Expert advice for the line organisation:

Development of the work community has become an important activity in which the OHS assists the **line organisation**. Fundia, in collaboration with FIOH, has created Metal-Age, which is a tool-box for participatory planning of tailored interventions at enterprise or team level, aiming at improved total productivity, better well-being and good work ability[2]. There are four steps in the Metal-Age planning:

1. Orientation phase
2. Intervention planning phase 1
3. Prioritizing phase
4. Intervention planning phase 2

The planning session starts with the Orientation Matrix (Figure 19.2).

ORIENTATION MATRIX

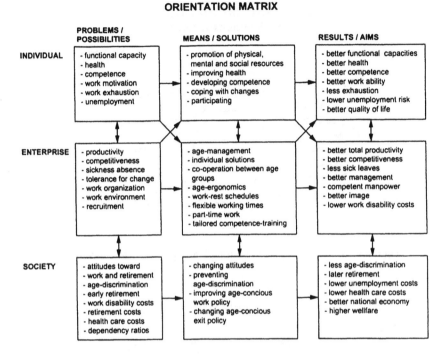

	PROBLEMS / POSSIBILITIES	MEANS / SOLUTIONS	RESULTS / AIMS
INDIVIDUAL	- functional capacity - health - competence - work motivation - work exhaustion - unemployment	- promotion of physical, mental and social resources - improving health - developing competence - coping with changes - participating	- better functional capacities - better health - better competence - better work ability - less exhaustion - lower unemployment risk - better quality of life
ENTERPRISE	- productivity - competitiveness - sickness absence - tolerance for change - work organization - work environment - recruitment	- age-management - individual solutions - co-operation between age groups - age-ergonomics - work-rest schedules - flexible working times - part-time work - tailored competence-training	- better total productivity - better competitiveness - less sick leaves - better management - competent manpower - better image - lower work disability costs
SOCIETY	- attitudes toward work and retirement - age-discrimination - early retirement - work disability costs - retirement costs - health care costs - dependency ratios	- changing attitudes - preventing age-discrimination - improving age-concious work policy - changing age-concious exit policy	- less age-discrimination - later retirement - lower unemployment costs - lower health care costs - better national economy - higher wellfare

Figure 19.2 The Metal-Age Orientation Matrix

During Intervention phase 1, the participants usually work in pairs and fill in suggestions for solutions in the ovals. The main aims are seen at the periphery of the central circle and in the middle is written the level at which the planning is done. The example is from a tube factory (Figure 19.3).

The next step is Prioritizing. This is done with the Prioritizing Matrix. The participants give points to each solution, starting with the importance of the problem for which the solution is planned. After that is the width, which indicates how many employees are concerned. Finally, points are given for the ease with which the solution can be implemented.

The example is from the same planning session as in Figure 19.3 (Figure 19.4).

Intervention Planning
(Aims-Solutions-Actions)

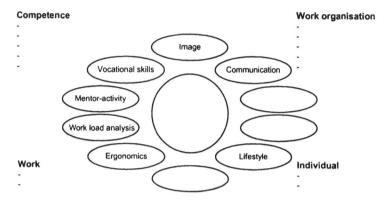

Figure 19.3 Metal-Age Intervention planning phase 1 at a tube factory

Priority Matrix

	Importance (1-10)	Width (1-10)	Ease (1-10)	Result (1-100)
Lifestyle	9	6	4	216
Ergonomics	10	6	7	**420**
Vocational skills	7	4	5	140
Communication	10	7	6	**420**
Image	8	8	5	320
Work load analysis	8	8	3	192
Mentor-activity	7	4	6	168

Figure 19.4 The Metal-Age Prioritizing phase from a tube factory.

Finally, the participants plan activities for developing Competence, Work organisation, Work and Individual for the solution that scored highest in the Prioritizing phase.

The example is for Ergonomics for the same tube factory as above (Figure 19.5).

Intervention Planning
(Aims-Solutions-Actions)

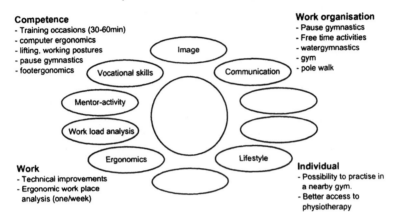

Competence
- Training occasions (30-60min)
- computer ergonomics
- lifting, working postures
- pause gymnastics
- footergonomics

Work organisation
- Pause gymnastics
- Free time activities
- watergymnastics
- gym
- pole walk

Work
- Technical improvements
- Ergonomic work place analysis (one/week)

Individual
- Possibility to practise in a nearby gym.
- Better access to physiotherapy

Image

Vocational skills

Communication

Mentor-activity

Work load analysis

Ergonomics

Lifestyle

Figure 19.5 Activities for improving Ergonomics at a tube factory.

One of the OHS nurses uses picture therapy or Metal-Age combined with picture therapy for development of the work community. The development of the work community can also concern a special topic defined earlier by the work community. Solution-based techniques commonly are used.

The **line organisation** mainly handles the development of work content and vocational skills.

In today's industry, supervisors experience a high level of stress. The annual day for supervisors is meant for education and discussions. The supervisors choose the topics themselves.

The Birthday discussion will be a miniature development discussion. The idea is to create a positive atmosphere over a cup of coffee by giving a small gift to every employee on his/her birthday. The employee's immediate superior will do this. Some structured questions will be asked every year.

DISCUSSION

The Fundia Wireys program is agreed on by both management and employees. Some parts of the program have not yet been implemented. The fundamental principles in Fundia Wireys are participation, comprehensiveness and continuity.

CONCLUSION

The experience of the Fundia Wireys program indicates that activities initiated by the OHS can improve the work ability and well-being of the employees and the total productivity of the company.

REFERENCES

1. Näsman, O., 1999, The Respect for the Aging Program at Fundia. In FinnAge-Respect for the Aging: Action Program to Promote, Health, Work Ability and Well-being of Aging Workers in 1990-96. In *People and Work, Research Reports 26*, edited by Ilmarinen, J. and Louhevaara, V., Finland: Finnish Institute of Occupational Health, Helsinki.
2. Näsman, O. and Ilmarinen J., 1999, Metal-Age: A process for improving well-being and total productivity. In *Experimental Aging Research Vol. 25, Number 4*, USA.
3. Näsman, O. and Ahonen G., 1999, The DalBo-project: The economics of maintenance of work ability. In *Aging and Work 4*, edited by Willem J.A. Goedhard, Healthy and Productive Aging of Older Employees, The Hague, The Netherlands.

20. Changes in the Work Ability Index of Aging Workers Related to Participation in Activities for Promoting Health and Work Ability: A 3-Year Program

Veikko Louhevaara[1], Anneli Leppänen[2], Soili Klemola[2]

[1] Kuopio Regional Institute of Occupational Health and University of Kuopio, Kuopio, Finland
[2] Department of Physiology, Finnish Institute of Occupational Health, Helsinki, Finland

ABSTRACT

A multidisciplinary intervention of three years (the MAHIS program) for promoting health, work ability and well-being was carried out and studied in a Finnish factory of an international enterprise manufacturing electromechanical devices. The aim of this study was to examine the positive and negative changes of the work ability index (WAI) of aging workers related to participation in various activities completed within the MAHIS program. The subjects were 263 voluntary aging (45+ years) workers (194 men and 69 women). Within the MAHIS program, 16 types of training or cultural activities were organized for promoting health and work ability. The WAI of the 68 subjects increased more than two points during the program. The results suggest that the good rate of participation in various types of activities of the MAHIS program, and particularly, in the activities, which were done together with other workers, was associated with positive changes in the WAI of aging workers. It is also probable that the good rate of participation also reflected the general active life-style of these aging workers.

Keywords: *Work ability, Age, Health, Fitness, Promotion*

INTRODUCTION

In Finland, it was realized about 10 years ago that decreased workforce participation rates of older workers are a severe risk for the economy. It was also predicted that this problem would increase due to the aging of the baby boomers born in 1945-1950. The only short-term solution for survival was to promote the health, work ability and well-being of aging workers using all possible measures in all circumstances. Therefore, in many enterprises systematic, comprehensive and need-based programs were started to promote the work ability of workers. One of the programs that was scientifically planned, monitored and evaluated was the so-called MAHIS program. It was carried out in a Finnish factory of an international enterprise manufacturing electromechanical devices [1]. The need for initiating the MAHIS program was mainly based on the following changes in the working life: aging of workers, increasing demands for efficiency, and constant changes in the business environment.

The theoretical framework of the MAHIS program is based on a comprehensive and multidisciplinary concept for promoting health, work ability and well-being (Fig. 20.1) [2]. The concept was created and then continuously developed for practitioners of occupational health and safety according to the results of a longitudinal study of municipal workers, [3] and also based on experiences obtained from several studies of the Respect for the Aging - FinnAge - Action Programme [4]. According to the concept, four types of measures are needed to promote health, work ability and well-being of workers. The measures are focusing on the work content and work environment; work organization; individual resources; and professional competence. In the work and environment, for example, load factors and various exposures can be decreased by ergonomic, safety and industrial hygienic measures. In work organization, management relations and relationships between workers can be improved and more flexible work times arranged. At the individual level, such possibilities as participating in regular exercise and various cultural activities can be supported. The maintenance of high-qualified professional competence requires continuous learning of new knowledge and skills, i.e., a life-long learning process. As a consequence of better health and work ability, the productivity and quality of work as well as perceived well-being at the individual and organization levels are improved. All preventive measures completed during an occupational career affect the "Third Age," which can be meaningful and independent of health and social services for many years.

The 3-year MAHIS program included an initial survey done in 1995, and such factors as the health status, work ability, functional capacity, and work load of approximately 800 workers were assessed. After the various tailored and need-based measures that were carried out within the MAHIS program, the same assessments were completed in the final survey in 1998. Effects and feasibility of the 3-year MAHIS program were studied using a multidisciplinary approach [1].

The aim of the present study was to examine the positive and negative changes in the work ability index (WAI) of aging workers related to participation in various activities at the individual level completed within the MAHIS program.

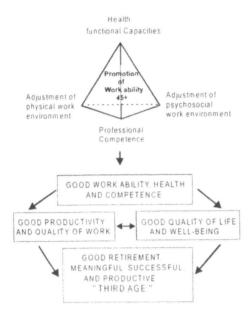

Figure 20.1 Basic concept for promoting work ability and health during aging and it's consequences on productivity of work, workers' well-being and "Third Age"

MATERIAL AND METHODS

Subjects

The number of workers who participated voluntarily in the initial and final assessments of the MAHIS program was 772 men and women. The subjects of this study on changes in the WAI were 263 aging (45+ years) workers (194 men and 69 women) who participated in the assessments of health and work ability both before and after the 3-year MAHIS program.

The subjects were divided into two groups according to the changes in the WAI during the MAHIS program: those whose individual WAI decreased more than 2 points (85 subjects, the negative group) and those whose individual WAI increased more that 2 points (68 subjects, the positive group). The mean ages for the positive and negative groups were 49 (SD 3) years and 50 (SD 4) years, respectively. The mean increase and decrease in the WAI in the two groups was significant when the values obtained in 1995 and 1998 were compared (p<0.001) (Table 20.2). The significant differences ensured that the goal of the study design was attained.

MAHIS program

The realization of the MAHIS program was based on so-called MAHIS groups. Each profit unit of the factory had its own group (a total of 13 groups), which were responsible for such factors as the application of the measures agreed upon and the participation of personnel.

The measures of each MAHIS group were planned according to the results of the initial survey. Feedback from the results was given to the workers at seminars in which work-related ergonomic and other improvements were also planned. Suggestions for ergonomic improvements were taken back to the various profit units for realization. In the development of the work community, special attention was paid to supervisor and team training. In order to strengthen functional capacity, different exercise groups were established.

Within the MAHIS program and its groups, 16 types of activities were organized and completed for promoting health, work ability and well-being as follows:

- non-specific training including various physical activities and events
- guided water gymnastics
- guided indoor gymnastics
- pause gymnastics
- playing various ball games such volley and basketball
- short trips by, for instance, bicycle or boat
- long trips, for instance, to Lapland
- fitness tests
- health examinations
- fitness card campaigns
- MAHIS seminars during weekends outside of the factory
- cultural activities such as theatre-going
- company excursions and excursions to fairs
- training in team work
- training for improving supervisors' leadership skills
- training for controlling and alleviating work-related stress

Methods

The WAI, i.e., perceived work ability according to the work ability index, was measured by a questionnaire-based index formed from seven items (Table 20.1) [5]. The final index score ranges from 7 to 49, and is divided into four perceived work ability categories as follows: poor (7-27 points); moderate (28-36 points); good (37-43 points), and excellent (44-49 points).

Statistical analysis

The differences between the compared groups in terms of the WAI and the participation rate of the subjects in various activities were tested with the t-test for paired observation and the Fisher's Exact Test based on the Chi-Square-test. The tests were done using Version 6 of the SAS/STAT Package. The differences were considered significant when $p < 0.05$.

Table 20.1 Work ability index (WAI)

Item[a]	Scale (points)
(1) Subjective estimation of current work ability compared with lifetime best	1-10
(2) Subjective work ability in relation to both physical and mental demands of the work	2-10
(3) Number of diagnosed diseases	1-7
(4) Subjective estimation of work impairment due to diseases	1-6
(5) Sickness absenteeism during the past year	1-5
(6) Own prognosis of work ability after two years	1,4,7
(7) Psychological resources (enjoying daily tasks, activity and life spirit, optimistic about the future)	1-4

[a] The items are explained as follows: (1) 1 = very poor, 10 = very good; (2) 2 = very poor, 10 = very good; (3) 1 = five or more diseases, 7 = no diseases; (4) 1 = fully impaired, 6 = no impairment; (5) 1 = 100 days or more, 5 = zero days; (6) 1 = hardly able to work, 4 = not sure, 7 = fairly sure; and (7) 1 = very poor, 4 = very good.

RESULTS

The mean WAI of the positive group was 4.0 points higher ($p < 0.001$) in 1998 than in 1995. Correspondingly, the mean WAI of the negative group was 5.2 points lower ($p < 0.001$) (Table 20.2).

Table 20.2 The work ability index (WAI) of the male and female subjects in the positiv group (individual WAI increased more than 2 points) and in the negative group (individual WAI decreased more than 2 points) in 1995 and 1998

	N	1995 Mean	SD	1998 Mean	SD
Men and women					
Positive	68	38.3	6.1	42.3***	5.5
Negative	85	41.8	5.6	36.6***	6.6

*** $p < 0.001$

The positive group participated more than the negative group in the activities that were organized within the MAHIS program (Table 20.3).

Table 20.3 The proportion of the male and female subjects (%) in the positive group (n=68) ands in the negative group (n=85) who participated in all or almost all occasions or events organized within the 16 activities

Activity	Positive Mean (%)	Negative Mean (%)	Difference Mean (%)
Non-specific training	70	59	11
Guided water gymnastics	14	13	1
Guided indoor gymnastics	9	8	1
Pause gymnastics	38	31	7
Playing various ball games	3	1	2
Short trips	42	39	3
Long trips	17	16	1
Fitness tests	70	62	8
Health examinations	97	99	-2
Fitness card Campaigns	54	60	-6
MAHIS seminars	83	72	11
Cultural activities	54	30	24**
Excursions	45	44	1
Training in team work	80	84	-4
Training for supervisor work	23	19	4
Training for work-related stress	25	14	11

** $p<0.01$

In the positive group, the mean relative proportion of the subjects who participated in all or almost all occasions and events was significantly greater than that in the negative group within the cultural activities (24%, $p<0.01$). A similar quite strong trend with the mean difference of 11% could be seen in the following activities: non-specific training, participation in the MAHIS seminars, and training for controlling and alleviating work-related stress (Table 20.3).

DISCUSSION

According to the concept for promoting health, work ability and well-being (Fig. 20.1), changes in the WAI follow changes in individual characteristics and life style, work and the work environment, work organization and professional competence. The longitudinal study of aging municipal workers showed that both improvement and worsening of the WAI were significantly associated with changes at work and in individual life style and behavior. A model of the improved WAI included increased satisfaction with the supervisor's attitude, fewer repetitive work movements, and increased vigorous physical exercise during leisure time. Therefore, improvements in occupational ergonomics, management (leadership) and physical activity promote perceived work ability [3]. Later, based on the experiences obtained from the studies in the Respect for the Aging - FinnAge - Action Programme [4], the dimension of professional competence was observed to be very important in terms of older workers' work ability and well-being, and was added to the concept.

In the study by Soininen & Louhevaara [6], policemen (N = 505) participated in a 3-year program for promoting health and work ability. During the program, physical activity and fitness increased and cardiovascular risk factors decreased. The stress index increased but no changes were observed in the index of work satisfaction, work stress and satisfaction. The average WAI decreased significantly. However, there were 108 subjects whose WAI increased more than two points during the program. The positive changes in the WAI were significantly related to perceived lower stress and increased work satisfaction.

The present study supports the previous results [3,6]. The positive changes in the WAI were associated with the participation rate in such activities that were organized in groups. The good participation rate in various activities organized within the MAHIS program might also reflect a general active life-style.

CONCLUSIONS

The results of this study suggest that the good rate of participation in various types of activities of the MAHIS program, and particularly in the activities that were done together with other workers, was associated with the positive changes in the WAI of aging workers. The good rate of participation probably also reflected the general active life-style of these aging workers.

REFERENCES

1. Ilmarinen, J., 1999, Ageing workers in the European Union-Status and promotion of work ability, employability and employment, Finnish Institute of Occupational Health, Ministry of Social Affairs and Health, Ministry of Labour, Helsinki, pp. 243-246.
2. Ilmarinen, J., Louhevaara, V., Huuhtanen, P. and Näsman, O., 1999, Developing and testing models and concepts to promote work ability during ageing. In *FinnAge-respect for the aging: Action programme to promote health, work ability and well-being of aging workers in 1990-1996*, edited by Ilmarinen, J. and Louhevaara, V., People and work. Research reports **26**, Finnish Institute of Occupational Health, Helsinki, pp. 263-267.
3. Tuomi, K., 1997, Eleven-year follow-up of aging workers. In *Scan J Work Environ Health 23 suppl 1*, pp.71.
4. Ilmarinen, J. and Louhevaara, V., FinnAge-respect for the aging. In *Action programme to promote health, work ability and well-being of aging workers in 1990-1996*, People and work. Research reports 26, Finnish Institute of Occupational Health, Helsinki.

5. Tuomi, K., Ilmarinen, J., Jahkola, A., Katajarinne, L. and Tulkki, A., 1998, Work Ability Index. In *Occupational Health care 19*, Finnish Institute of Occupational Health, Helsinki.
6. Soininen, H. and Louhevaara, V., 2000, factors associated with changes of the work ability index in policemen: a three-year program. In *Proceedings of the IEA 2000/HFES 2000 Congress*, 2000 July 29-Aug 4, Human factors and Ergonomics Society, San Diego, Santa Monica, pp. 172-174.

21. The Strict Agricultural Products Standard and the Difficulty of Agricultural Work for Aged Workers

Yoshie Shimodaira[1], Nobuo Ohashi[2]

[1] Department of Life Science, Nagano Prefectural college, Nagano, Japan
[2] Faculty of Social and Information Sciences, Nihon Fukushi University, Aichi, Japan

ABSTRACT

In Japan it is estimated that 50% of agricultural workers will be aged people in 2010[1]. With the aging of agricultural workers, abandonment of cultivated land has increased, and agricultural productivity has declined. Finding what is necessary for aged workers to continue their agricultural work is a very important issue today, because if aged workers can continue agriculture, it is expected that the poor food self-sufficiency rate in Japan could be improved. There are many problems to be solved concerning agriculture for aged workers. One concerns the present agricultural products standard. We discovered this fact through our interview and observation study about the life of aged people living in the village of Nakajo in Nagano Prefecture, where depopulation has been advancing. There is a standard that defines characteristics such as size, shape, weight, etc. of agricultural products for supply to the market. Since the present standard is too strict for aged workers, they are easily eliminated from the cultivation of the agricultural products for supply to the market, in spite of having the capacity to cultivate various foods. We propose the easing of the agricultural standard in order to maintain work capability of aged agricultural workers.

Keywords: *Aging, Agriculture, Products standard, Work capability*

INTRODUCTION

The aging of Japanese society has had a major effect on agriculture and has brought aging of agricultural workers, a decrease in the number of agricultural workers, and reduction of farmland. For example, although the number of agricultural workers was 10,250,000 in 1970, it has dropped continuously since reaching 4,140,000 in 1995. Among these, although there are 2,560,000 commercial agricultural workers, this number will fall to 1,470,000 in 2010. Moreover, as the aging of commercial agricultural workers has progressed, the number of those aged 65 or older has risen. The figure was 17% in 1980, but rose to 40% in 1995, and is estimated to reach 50% in 2010 [1].

As commercial agricultural workers age, they generally stop commercial agriculture and move to agriculture for their own consumption. As a result, farmland decreases. In particular, abandoned cultivated land in mountain areas is increasing [2]. On the other hand, with the progress of industrialization, the food self-sufficiency rate in Japan has continued to fall every year, and what was 73% in 1965 has fallen to 40% in 1999 [3]. While it is expected that food supply and demand in the world are tight on a mid- and long-term basis, it is recently thought increasingly that an improvement in the self-sufficiency rate is a major issue facing Japanese society. Meanwhile, it is necessary to maintain the work capability of aged agricultural workers, and to improve the situation so that they can continue not only self-consumption-agriculture but commercial agriculture that supplies agricultural products to the market. This will also contribute to an improvement in the food self-sufficiency rate. Based on our research carried out in the village of Nakajo, we discuss in this paper the measures required for aged workers to continue commercial agriculture.

AGRICULTURE OF VILLAGE NAKAJO

Since 1989 we have carried out research about the life of aged people living alone in the depopulated mountain village of Nakajo. Aging progresses here every year and the present ratio of elderly people is 38.5%. The number of households is 1,037 houses, among which the number of farmhouses is 529 houses, which is 51.4%. The number of agricultural workers (commercial farmhouses) is 262, which is 9.2% of the population of this village [4]. Most of the aged workers are growing agricultural products for their own consumption. 72.5% of commercial agricultural workers are aged people. However, very few aged people who are living alone are supplying products to a market. Even if they do so, the quantity and income are negligible [5]. Those who have continued supply for a long period stop when they become aged workers.

Our research found that there were three major factors in stopping commercial agriculture that supplies agricultural products to the market. The three factors are: the decline in physical strength, difficulty in using machines, and difficulty in satisfying the rigorous agricultural products standard.

1. Decline in Physical Strength

As physical strength declines with aging, generally the supply of heavy vegetables whose weight exceeds 2kg per one piece, such as pumpkin, Chinese cabbage, and watermelon, will become difficult. Even if cultivation is possible for self-consumption, commercial agriculture is difficult because there are various processes that require physical strength, such as harvesting, sorting, washing, packing, and carrying. Therefore, generally agriculture changes as follows with aging. The first change accompanying aging is to change the agricultural products grown to lighter vegetables from heavy vegetables. Elderly people can continue commercial agriculture by adopting such a method. However, if supply becomes impossible even with this method, they will cultivate various vegetables for self-consumption. Then, aged people will stop the cultivation of heavy vegetables for commercial agriculture, and finally will completely stop agriculture.

2. Difficulty in using machines

Since the whole village is located in a mountain area, almost all farmland is located on inclined land. There is even a place where the degree of inclination exceeds 30 degrees. Due to this land situation, it is difficult for aged people to use farming machinery. Moreover, there are many elderly people who cannot start the engine of machines, or who cannot carry machines from the house to the field [6]. This makes the burden of agricultural work greater for aged people.

3. Difficulty in satisfying the rigorous agricultural Standard

This is one of the most important points that we discuss in this paper.

Aged people who can continue self-consumption agriculture can barely participate in commercial agriculture. Because the present agricultural product standard is too rigorous for aged workers, they have difficulties in supplying products that meet the standard. Although elderly people have the capability to grow agricultural products that can be used as food, they are eliminated from the market.

Agricultural products are treated just like industrial products. The standard consists mainly of the grace standard in connection with the color and shape of appearance, the size standard in connection with size and weight, the packing standard in connection with the size of the packing container and the packing method. The present standard is defined aiming to raise the price of goods by improving appearance, to increase the efficiency of transportation, and to reduce bruising of agricultural products during transportation. There is an agricultural product standard that is defined by The All Japan Agricultural Cooperative Association; however, a more rigorous standard competes and is defined by local agricultural cooperative associations. For example, although the bend of a cucumber which is less than 2cm is Class A by the All Japan standard, the Nagano Agricultural Cooperative Association, which is a major production location for summer vegetables, has determined that a Class A article has a bend of less than 1cm [7]. Because an association takes charge of collection and supply of agricultural products, refused products cannot be supplied to the market as food, even if such products essentially are useful as food.

EXAMPLE OF AGRICULTURAL STANDARD

The Nagano Agricultural Cooperative Association defines the agricultural standard for 38 kinds of vegetables, five kinds of mushrooms, and 13 kinds of fruits. The example of a standard is shown below.

Cucumber

The standard for a cucumber is divided into six categories by bend and size. A cucumber with a bend of less than 1cm is classified as Class A, and is further divided into three categories by size. A cucumber with a bend of over 1cm and less than 2cm is classified as Class B, and is further divided into two categories by size. The packing standard specifies that one box should weigh 5kg.

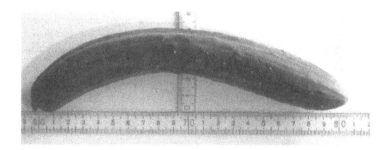

Figure 21.1 Cucumber with bend of 2.0cm or more (Class C)

Table 21.1 Products standard for a cucumber

Grade class	Size class	Weight (g)	Length (cm)	Number of one carton	Grace standard
A	S	90-100	18-20	51-57	Quality and color is good, clean and fresh, no ill or noxious insects, no scratch, the bend: less than 1.0cm, the length: almost same.
	M	100-120	20-22	43-50	
	L	120-140	22-24	37-42	
B	S	90-100	18-24	51-57	Applying to Class A correspondingly, the bend: less than 2.0cm.
	M	100-140	20-24	37-50	
C	M	90-140	18-24	37-57	Applying to Class B correspondingly, the bend: less than 3.0cm.

Source: Agricultural Product Standards of Vegetables-and-Fruits, Nagano Prefecture, 1999

Mini tomato

Mini tomatoes are divided by weight into five categories: 2L, L, M, S, and 2S. The packing standard requires one package to weigh 200 gr. In order to meet this packing standard, the number of mini tomatoes put into one pack by size must change. Size 2L needs 10 or fewer pieces. Size L needs 11 to 15 pieces. Size M needs 16 to 22 pieces, Size S needs 23 to 30 pieces, and Size 2S needs 31 or more pieces. Next, 18 packs containing the same size tomatoes are placed into one carton. The weight of one carton thus becomes 3.6kg.

Plum

Plum are divided into six categories: M, L, and 2L, 3L, 4L and 5L, by size. The conventional sorting device currently used at each farmhouse can sort three sizes only. On the other hand, dividing into six categories is demanded by the present standard. Then the collecting center classifies collected plums of three kinds into six kinds using a new machine at the center.

These are some examples of the agricultural products standard. Although these standards are to be improved every two years, a new standard for a new category of article may be defined or a weight standard may be changed; however, the grace standard and size standard are seldom changed.

EXAMPLE OF AGRICULTURAL WORK OF AGED PEOPLE

Thus, in order to sell products satisfying an agricultural standard defined very rigorously for aged people, much hard work is needed. Cultivating, harvesting, sorting, packing, carrying and so on create a large burden for aged people.

Here we describe the work of elderly people who are supplying cucumbers, mini tomatoes, and plums to the market.

1. Cucumber Harvesting Work

Mr. S and his wife are both 59 years old and have been producing cucumbers. They get up at 4:00 in the morning and start farming. In order to produce cucumbers that do not bend, the husband and his wife must assemble pillars supporting a halm and frequently bind the halm to a net. They must check that the halm does not fall over due to the weight of the cucumber. Also, they must make sure that cucumbers touch neither halm nor supporting pillar. Moreover, they must disinfect every 10 days to keep the cucumbers healthy. Harvesting usually takes place twice a day, morning and evening. However, in the season when the temperatures are highest, daytime harvesting also is necessary. Otherwise, the size of the cucumber becomes too large to meet the standard in a very short time. They cannot rest from work during harvest time, from the beginning of July to the end of October. During this period they usually can sleep only four hours a day, so they are often dozing unawares while they are sorting cucumbers at midnight. Therefore, the number of farmhouses that supply cucumbers to the market is decreasing. Nowadays only two such houses still remain in the village.

Mrs. N, an 80-year-old female farmer, stopped supplying cucumbers 15 years ago and started to supply mini tomatoes. "I was about to be killed by cucumbers," she said [8]. Her words illustrate how great the burden is on farmers who supply cucumbers for the market.

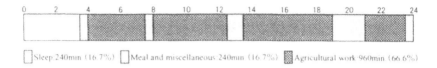

Figure 21.2 One day of Mrs. S. in the highest season of the harvest of a cucumber

Figure 21.3 Cucumber harvesting work

2. Sorting Work

After harvesting farm products, sorting work to meet the standard is required.

This work is also very hard. We describe the example of sorting work for plums, one of the typical agricultural products of the village.

Plums are harvested once a year. The period for supply is determined as ten days in early June. During these ten days, harvesting and supply must be completed. So harvesting is performed in the daytime and sorting is performed under electric light from sunset to midnight. Removing foreign substances and leaves, the plum farmer divides the fruit into three size categories using a simple sorter to meet the size standard. The farmer then leaves the plums until the next morning. She gets up at 4:00 a.m. and starts to check again to remove plums that have small cracks or small stains so that the plums meet the grace standard. As plum sorting work deals with imperfections of a very small size and generally is done under light of insufficient brightness, the burden of this work is very great, particularly for aged people whose visible acuity has already declined.

Table 21.2 shows an example of one day in the life of an elderly female farmer, Mrs. N, during the harvesting period. Her working time was 810 minutes and the time for meals and miscellaneous was 300 minutes, while sleeping time was only 330 minutes. Since such a life continues for ten days, the burden is very great for elderly people.

Figure 21.4 Plum sorting work

Table 21.2 One day in the life of Mrs. N, a plum farmer

Action	time	%
Sleep	330	22.9
Meal and Miscellaneous	300	20.8
Agricultural work	810	56.3
Total	1440	100.0

3. Packing Work

Packing work is also very hard. We describe the example of packing work for mini tomatoes. A mini tomato is one of the farm products that aged people can cultivate relatively easily; however, packing them for supply requires a great deal of time. Because the packing standard for mini tomatoes is very rigorous, as mentioned above, the farmer has to measure one by one, using a scale for every pack, and this takes a long time. In order to complete packing of all mini tomatoes harvested from 500 trees in the morning, a great deal of time is required in the afternoon. Sometimes, during the peak of the harvest season, the packing of all the tomatoes harvested in a day is not completed within the day. In such a case, the farmer resumes work from dawn of the next day. The harvest period of the mini tomato continues from around July to around October.

Although the packing of mini tomatoes must be done by hand and is very time-consuming, the weight of one box is 3.6kg. Elderly people are sufficiently able to manage boxes of this weight.

Figure 21.5 Mini tomato packing work

However, regrettably there is a packing standard that defines a heavy weight standard which is a burden for elderly people. For example, the weight of one box of Chinese cabbage is 15kg, and of onions, watermelons, or others is 20kg.

4. Carrying Work

After the packing is finished, the products must be carried. Most aged people use a wheelbarrow to carry products, because they do not have automobiles anymore. One aged farmer loaded four boxes of cabbage and carried them to the product collection location about 300 meters away from his house. As the weight of one box was 10kg, it meant that he carried cabbage weighing 40kg. He repeated this several times in a morning. As this example shows, carrying work places a very large muscular burden on elderly people who do not have an automobile. The farther away the product collection location, the more muscular burden is required. If an aged farmer is lucky enough, a neighbor who has an automobile may help him to carry his products. However, if carrying work becomes impossible for aged people, they generally have to give up supplying products to the market.

As mentioned above, if an aged farmer wants to supply agricultural products to the market, he carry out various tasks such as harvesting, sorting, packing, and carrying, in addition to cultivation. Each task is necessary in order to meet the severe agricultural standard, but each task also means a very large burden for aged people. If an aged farmer finds it difficult to carry out one of these tasks, he has to stop supplying products to the market.

PRACTICAL USE OF THE WORK CAPABILITY OF AGED AGRICULTURAL WORKERS

The main reason that an aged agricultural worker changes to self-consumption agriculture from commercial agriculture is the burden of the various tasks required to meet the severe agricultural products standard. However, aged farmers who are engaging in self-consumption agriculture have the capability to produce foods that can be used fully.

We must improve the present situation whereby the working capability required to produce foods causes the producers to be eliminated from the market. The situation stems mainly from the severity of the agricultural standard. Various methods to utilize the work capability of aged agricultural workers can be considered. In this paper we stress the alleviation of the agricultural products standard, the development of species that aged people can produce easily, and the introduction of work support systems.

1. Alleviation of Agricultural Products Standard

We have described how the severity of the present agricultural product standard means that very hard work is required to supply products to the market. We also described how these tasks create a heavy burden, particularly for aged farmers.

If the standard can be eased, it can be expected that the number of elderly people who can supply products to the market will increase. It is especially important to ease the grace standard, size standard, and packing standard. This is because the burden of cultivation work, sorting work, and carrying work will be mitigated by such easing.

However, there is one potential problem in doing so. If the standard is eased, for example, mini tomatoes of various sizes and bent cucumbers will appear in the shops. Some consumers may not accept such vegetables at first. However, prudent consumers soon will learn to purchase these vegetables. This is because the consciousness on the part of consumers relative to agricultural products has been changing. According to the results of an investigation performed by the Tokyo local government about consumers' purchasing awareness, consumers expect farmers to produce safe food that uses few agricultural chemicals. They also expect merchants to sell fresh vegetables, even if the appearance is not graceful, like the curved cucumber [9]. These were the top two expectations.

Another example is observed in the shop managed directly by the village of Nakajo. It is a fact that many tourists are willing to buy agricultural products that do not meet the standard, and they also are pleased with processed foods made from agricultural products that do not meet the standard. These facts mean that consumer awareness relative to agricultural products has been changing. Nowadays what the consumer is seeking is not graceful vegetables, but fresh and safe vegetables.

In the future, the aging of agricultural workers is unavoidable. On the other hand, consumers are looking for safe and fresh vegetables. The agricultural productive capacity as the work capability of aged people should be maintained. As one method of realizing this, it is necessary to alleviate the severity of the agricultural standard, on the basis of cooperation between agricultural cooperative associations and merchants, in addition to the understanding of the consumers.

2. Improvement of Species of Vegetables That Aged People Can Produce Easily

It is also effective to change the weight of agricultural products in order to maintain aged people's agricultural productive capability. Trials already have been carried out for some agricultural products. For example, there is a new species of Japanese pumpkin that is lighter and smaller. As this new pumpkin can sit a man's palm and the weight of one piece is about 400g (about 1/4 of the native species), aged people can handle them easily. This new pumpkin has been produced from 2000. These improvements bring benefits not only to aged workers, but also to consumers, who can use such vegetables more easily and conveniently. Therefore, the number of aged people producing them is growing little by little. This was observed in the village of Nakajo.

In an aging society, it is important to produce vegetables of smaller size and lighter weight. This is useful for both consumers and producers. Therefore, improvement of the species is required for such heavy vegetables as cabbage, Chinese cabbage and daikon radish, as well as the pumpkin mentioned above.

3. Introduction of Agricultural Work Support Systems

The burden of agriculture changes with the seasons. The big problem for aged farmers who want to continue agriculture is carrying out tasks when the burden becomes temporarily larger.

For example, the big task to be performed first every year in the early spring is to dig up the ground that was covered by the snow and hardened during the winter. As most aged people have difficulty in using agricultural machines, this digging work is very hard. There are some who give up agriculture due to their lack of confidence to perform this task. In some types of agriculture, harvesting brings the biggest burden, because the harvesting period is short. For some, sorting brings the biggest burden, because the grace and size standards are rigorous. Aged farmers, especially aged people living alone, must perform all these agricultural tasks by themselves.

If at least one of these agricultural tasks becomes impossible, aged people cannot but stop agriculture. If support systems to perform some particularly hard work for aged people can be established, aged people can continue commercial agriculture to supply products to the market. We believe that this could be a project for NPO or some similar organization. Such a support system may also be effective to bring aged people who have already stopped selling farm products back into commercial agriculture.

CONCLUSION

There are many aged farmers who have the capability to produce agricultural products that can be fully used as foods. However, due to the present rigorous agricultural product standard, aged farmers are eliminated from the market. Therefore, it is necessary to alleviate the present standard in order to use the work capability of aged people. Since agricultural workers' aging continues to progress, this alleviation is a matter of urgency.

Moreover, since circulation of agricultural products is becoming international, alleviation of agricultural product standards also requires international cooperation.

We should abolish the method of defining standards for agricultural products according to a concept similar to that for industrial products.

It is also necessary to make efforts to establish support systems for agriculture by aged people, and to develop species of vegetables that are easy to produce.

The work capability of aged agricultural workers would be further utilized through these measures.

REFERENCES

1. Agricultural census, 1995.
2. *The increase in an abandoned cultivated land*, 1995, Ministry of Agriculture, Forestry, and Fisheries The Kanto agricultural administration office.
3. *Food self-sufficiency rate report*, 1999, Ministry of Agriculture, Forestry, and Fisheries.
4. Heisei 12 editions. In The index of cities, towns and villages exception 100 classified by Nagano area, 2001, The Nagano plan office information aid policy division.
5. Shimodaira, Y. and Ohashi, N., 1995, Real Situation of Agricultural Work of Aged Persons living alone in a Depopulated Area and the Denudation of their Farm Land. In *Journal of Nagano Prefectural College*, **No. 50, 77**, pp. 71-89.
6. Shimodaira, Y. and Ohashi, N., 1995, Agricultural work of Aged Persons in a Depopulated Village in Highland-In Cases of Aged Persons Living Alone without Family in Village A-. In *The Paths to Productive Aging*, edited by Masaharu, K., (Taylor & Francis), **20**, pp. 18-22.
7. Agricultural product standards of vegetables-and-fruits, such as the Nagano Prefecture, 1995 & 1999.
8. Shimodaira, Y. and Ohashi, N., 1990, Agricultural work of Aged Persons in Depopulation Area-In case of The Village A in Nagano Prefecture-. In *Journal of Nagano Prefectural College*, **No.45, 81**, pp. 77-88.
9. Tokyo consumption habits monitor questionnaire, Result of an investigation "Vegetables", (Tokyo life culture office), 1999.

22. Survey of Prospects for Elderly Care Workers

Hisao Nagata[1], Sunyoung Lee[2]

[1] *National Institute of Industrial Safety, Tokyo, Japan*
[2] *Department of Architecture, Chinju National University, Kyongnam, Korea*

ABSTRACT

In an effort to study conditions of care work in nursing homes for the elderly in Japan and determine the outlook for future elderly care workers, questionnaires were sent in March 1999 to 2000 nursing homes, of which 969 responded, enabling the following conclusion: under the current status of elderly care services in nursing homes where approximately half of the inmates have paralysis and dementia, persons aged 60 or more find it very hard to act as care workers, mainly because of a lack of support measures for heavy manual lifting as in bathing care and bed-to-wheelchair transfer. As discovered from personal opinions in the questionnaires, aged care workers are deemed well suited for mental care work based on their understanding of the needs of senior citizens requiring nursing care.

Keywords: *Care labour, Elderly care worker, Nursing homes*

INTRODUCTION

The oldest bracket of seniors in Japan, over age 75, accounted for 7.0 percent (8.88 million) of the nation's population in 2000. It is anticipated that the number will more than double to 15.6 percent (18.89 million) by 2025 [1]. Added to the advent of an aged society, the number of physically handicapped, dementia sufferers and bedridden cases among the elderly is estimated to have topped 2.8 million in 2000 and to reach as high as 5.2 million by 2025 [2]. Conversely, Japan's labour force, the 20 ~ 59 age segment, is predicted to dwindle by 8.71 million from 57.16 million in 2000 to 48.45 million by 2025 [1]. Thus, while the demand for aged care workers will rise sharply, we inevitably will face a dire shortage of labour in this sphere of endeavor.

The purpose of this survey is to investigate the current status of care work in nursing homes and consider the possibility of future care efforts by the elderly. Focusing on certain nursing homes that have played a major role among facilities for senior citizens and those disabled by paralysis or dementia, we endeavored to understand the state of care work today and look into the possibilities for care workers aged over 60.

METHOD

One questionnaire was sent to the manager of each nursing home. We asked the manager to choose an experienced senior housemother (or father) as a respondent. The survey was conducted 18th through 31st March, 1999. To avoid regional deflection, 2,000 facilities were selected at random from among 3,976 elderly nursing homes listed in the directory of Facilities for Seniors and the Disabled [3]. As a result, 969 responses (48.5 percent) were obtained. Items requiring replies included:

-   Respondent data (age, sex, years of service)
-   Facility outline (number of inmates, their level of self-sustaining capacity, extent of disability, and number of staff workers)
-   Current status of care work (age structure of workers, tasks requiring physical effort, tasks incurring mental stress, state of night shift care work)
-   Care work by the elderly (age limit, problems encountered working with them, matters requiring improvement to accept elderly care workers, advantage or disadvantage of employing elderly care workers)
-   Comments on care work in general

RESULTS

Of the 969 respondents, 204 were male (average age: 35.0), 761 were female (average age: 46.6), and 4 unknown. As Table 22.1 shows, about 80 percent of the respondents were women, many in their 40s and 50s (average age: 44.2). These two age groups account for almost three-fourths of the total respondents. Some 30 percent had five to nine years of service, about 40 percent had served ten to nineteen years, with the average length being 10.6 years. Table 22.2 shows the survey results on the ratio of paralysis and dementia of inmates. Table 22.3 shows the survey results concerning the self-sustaining ability according to bathing, excretion and transfer. About half had dementia and paralysis, while 47 ~ 59 percent of inmates required total care.

Table 22.1 Character of respondents

Gender					
	Male	Female			
	21.1	78.9 (%)			
Age					
20-29	30-39	40-49	50-59	60- (yr)	Average age
15.2	15.7	29.7	37.2	2.2 (%)	44.2 (yr)
Experience					
1-4	5-9	10-19	20-29	30- (yrs)	Average years
17.8	31.1	39.7	10.9	0.5 (%)	10.6 (yrs)

Table 22.2 Ratio of disabled inmates

Paralysis of inmates	Dementia of inmates
44.9 (%)	67.5(%)

Table 22.3 Self-sustaining ability of inmates

Activities	No care	Partial care	Total care
Bathing	9.1	31.7	59.2 (%)
Excretion	23.7	22.2	54.1 (%)
Transfer	33.6	19.1	47.3 (%)

Table 22.4 shows the number of inmates and staff workers. 60.6 percent of nursing homes accommodated from 50 to 79 inmates, whereas 54.6 percent of nursing homes employed from 20 to 39 staff worker. The average number was 68.5 inmates and 37.6 staff. To learn the age structure of care workers at each facility, the poll sought their number by age bracket (19 and under, 20s, 30s, 40s, 50s, 60s and over), volunteers excluded. The results were classified by the number of workers (0, 1- 4, 5-9, 10 and over) to obtain the composition by age groups as shown in Table 22.5. Facilities that do not have workers 19 years old and under account for about 87.4 percent, while those not retaining workers over age 60 account for around three fourths. Even at facilities that employ workers over 60, their number is very small: the total of "5 to 9" and "10 and over" accounts for a mere 4.2 percent. The results make it clear that only a small number of facilities have elderly care workers over 60 on their payroll. At present, the proportion of care workers in their 20s is very high, whereas the numbers of 19 and under and 60 and over is negligible. It can be judged that this tends to relate to the increase of new facilities and the high ratio of young persons staffing them.

Table 22.4 Number of inmates and staff

Inmates					
Below 49	50-79	80-99	100-119	Over 120	Average
4.1	60.6	16.1	13.4	5.8 (%)	68.5 (inmates)
Staff					
Below 19	20-39	40-59	60-79	Over 80 (staff)	Average
7.1	54.6	30.7	5.7	1.9 (%)	37.6 (staffs)

Table 22.5 Ratio of care workers by age-groups

Age	Below 19	20-29	30-39	40-49	50-59	Over 60 (yrs)
No workers	87.4 (%)	1.5 (%)	5.8 (%)	5.2 (%)	17.5 (%)	67.3 (%)
1-4 workers	12.2	15.4	53.4	36.7	42.8	28.5
5-9	0.3	32.3	32.0	38.3	26.7	3.5
Over 10	0.1	50.8	8.8	19.8	13.0	0.7

Table 22.6 treats the night shift per month and ordinary working hours per night shift. Three, four and five per month accounts for 74.4 percent. The mean number is 3.6 times. All of 85.4 percent work long (16 or 17 hours) on the night shift. To the question "Can you take a nap on a fixed schedule?," 83.7 percent replied either "Always" or "Sometimes," while 7.3 percent answered "Not at all." As for napping time, "1 or 2 hours" and "2 or 3 hours" accounted for 72.9 percent.

Table 22.6 State of night shift care work

Working hours	Below 5	6-7	8-9	10-13	14-15	16-17	Over	18 (hrs)
	0.1	0.2	2.1	1.3	4.4	85.4	6.5	(%)
Number of night shift	0	1	2	3	4	5	6-9	Over 10
	8.8	5.3	6.5	12.3	38.0	24.1	4.9	0.1 (%)
Nap	Always		Sometimes		Not at all		Others or unclear	
	43.2		40.5		7.3		9.0 (%)	

To the question "Can you take meals during fixed recess hours?", 72.4 percent answered "Always," 22.3 percent replied "Sometimes," 1.8 percent stated "Never" and 3.5 percent didn't know. Asked about a margin for time to help patients gain self-reliance, 70 percent of the respondents said there was no margin for time.

In response to the query on care work requiring physical strain and causing fatigue, as Fig. 22.1 shows that the ratio of bathing care is the highest, accounting for 45.1 percent, followed by approximately a fourth each for excretion and transfer.

About a fifth of the responses regarding care work that causes heavy mental stress (Fig. 22.2) cited accidents by patients under care, such as falling, as their main worry. This was followed by care for the elderly with dementia, difficulty in furnishing enough face-to-face care, and collapse of rhythm in life and sleep owing to night duty.

When asked their opinion if the aging trend of care workers will continue, 41.6 percent thought it would, 33.3 percent didn't think so, 12.9 percent thought it would not change substantially, and 12.1 percent did not know. To the query about the age limit of care workers (Fig. 22.3), 73.4 percent thought it should be 55 or 60, while a scant 7.2 percent thought they should be allowed to work until 65. The total of "until 70 years old" and "at any age as long as they're healthy" accounted for only 11.6 percent.

When polled regarding the most serious decline in physical functions among care workers over 60, 19.7 percent stated weakness of legs and back, 13.6 percent pointed to a slowing of physical movement and failing agility, while 12.7 percent mentioned difficulty in reading small print.

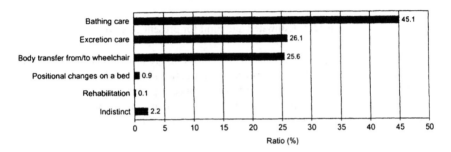

Figure 22.1 Tasks requiring physical efforts

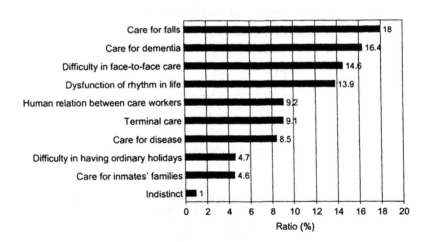

Figure 22.2 Tasks incurring mental stress

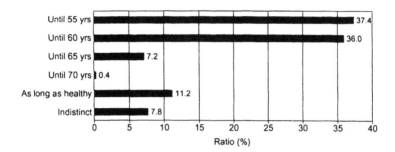

Figure 22.3 Estimated age limit for present care works

On the subject of problems when workers over 60 are teamed with their younger counterparts (Fig. 22.4), 21.2 percent cited moving patients between bed and wheelchair or stretcher, while 18.0 percent referred to bathing care, particularly transfer between the bath and transfer equipment. Both tasks are related to manual lifting. But approximately one-third thought there was no particular problem.

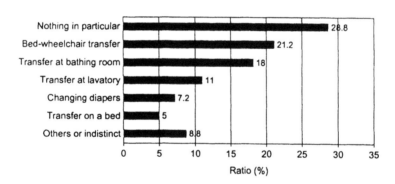

Figure 22.4 Problems encountered working with elderly care workers

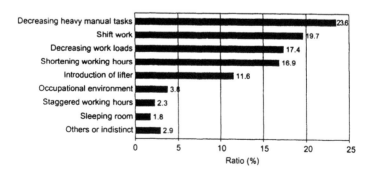

Figure 22.5 Necessary improvements to support elderly care workers

When asked what improvements would make the job of elderly care workers easier (Fig. 22.5), more than 20 percent responded that less work requiring muscle should do the trick, followed by a release from day plus night duty and night duty itself, reduced work-load and quotas, and shortened working hours, all aimed at making tasks for the elderly simpler.

A total of 395 respondents gave 543 opinions regarding the traits of aged care workers. As Table 7 shows, what they said can be roughly classified as:

- "Aged care workers understand the needs and position of persons requiring care (29.4 percent)"

Table 22.7 Advantages and disadvantages of employing elderly care workers

Advantages	
29.4 (%)	They can understand the state of mind or situation of aged inmates
23.0	They have abundant experience and knowledge.
20.1	They have topics of conversation familiar to aged inmates.
14.4	They can take better mental care of aged inmates.
13.1	They can provide a more easy and reliable atmosphere.
Disadvantages	
66.6 (%)	The physical burden is too heavy
10.0	Mental burden is too heavy
6.7	Weak eyesight
6.7	Deteriorated judgment
6.7	Unfamiliar with care equipment
3.3	Slow body-movements

- "They have abundant experience and knowledge (23.0 percent)"
- "They have topics of conversation familiar to aged inmates (20.1 percent)"

As Table 22.7 shows, 30 respondents cited problems related to elderly care workers, among whom 66.6 percent mentioned that

- "The physical burden is too heavy."

Regarding care work in general, 152 respondents gave their opinions, 53 of whom complained about the shortage of staff members, this accounting for about one-fourth of the total.

DISCUSSION

While the desire of Japan's senior citizens to work is extremely strong, the survey results show that nursing homes that do not have care workers over 50 years of age account for a fifth of the total, whereas facilities without care workers over 60 years old represent nearly 70 percent. Even at facilities that employ care workers over 60, they number fewer than four at each. As for the age limit of care workers, responses stating "possible until 65 years old" accounted for a mere 7.0 percent and "possible until 70" amounted to less than 1.0 percent, making it clear that the age limit among care workers is lower than in other lines of endeavor. Still, with respect to problems when care workers over 60 and their youthful counterparts work together, almost a third of the respondents said there was no special difficulty. Thus, the results do not deny the possibility of jobs for the elderly as care workers even now. The present working conditions at nursing care homes in Japan are severe, and an environment to accept elderly workers has not been fully established.

Accordingly, by building an environment conducive to hiring elderly workers instead of eliminating them, by introducing new concepts such as "Design for all", "Universal design" and "Real life design" in care support equipment, and reducing workload by cutting the amount of care tasks and working hours as well as making appropriate assignments, the scope of activities for elderly workers will expand. For this, it is essential to know the physical limitations of the elderly and their desire to participate in social activities.

REFERENCES

1. The Association of Employment Development for Senior Citizens, 2001, In *Handbook of Labour Statistics on Aged Society*, pp. 46-47 and pp. 56-57, (Japanese).
2. Ministry of Health and Welfare, 1998, In *White Paper on Health and Welfare in 1997*, (Gyousei Co. Ltd.), pp. 111, (Japanese).
3. Ministry of Health and Welfare, 1999, In *The Directory of Welfare Facilities for Seniors and Disabled in 1998*, Foundation of Social Development for Senior Citizens (Japanese).

23. A Program to Support and Maintain the Work Ability and Well-being of Kitchen Workers

Leila Hopsu, Anneli Leppänen, Soili Klemola

*Finnish Institute of Occupational Health, Department of Physiology,
Helsinki, Finland*

ABSTRACT

The work in school kitchens has changed dramatically during the past decade. The food-processing has been rationalized, the work tasks have been reorganized, and the productivity demands have increased. According to the personnel, the mean age of kitchen workers has risen and their work ability has decreased. In this study, a long-term program for maintaining the work ability of kitchen workers was created, evaluated and developed further. The study population included all the kitchen workers (n=198) in Espoo, the second largest city in Finland. The mean age of the workers was 44.8 years in 1997, and 98% of them were women. The three-year participatory intervention was started in 1997. It consisted of several activities supporting professional competence of employees'. The participants improved their work process in groups and participated in vocational training, as well as in a fitness program. The management also reorganized its own work and became acquainted with, e.g. age management. Both before and after the intervention, the Work Ability Index (WAI), the Occupational Stress Questionnaire (OSQ) and a measure of conceptual mastery of kitchen work were administrated to evaluate the effects of the program.

Keywords: *Aging, Kitchen work, Work characteristics, Well-being, Professional qualification, Maintaining work ability*

INTRODUCTION

The decreased rate of participation in the workforce among older workers is considered a serious economic risk for the national economy in several European countries. In Finland this trend has alarmed both politicians and enterprises for several reasons. The workforce participation rates in the older age groups have dropped dramatically during the past two decades. In 2000 the workforce participation rates was only 23% in the age group of 60-64 year old persons although the retirement age is 65 years. At the same time the mean age of the workforce will increase dramatically in the near future. The birth cohort born during the baby boom after World War II has just reached the age of 55 years. If the workforce participation rate continues to fall as in the older age groups, the national economy, or at least the pension level is in danger. Therefore, it is important to support the workers to maintain their work ability and motivation in all circumstances.

Some researchers have studied the prerequisites of maintaining work ability. The actions needed for maintaining the workers' work ability include four types of measures: (1) improving the work itself and the work environment, (2) developing the organizational culture of the workplace, and (3) promoting the health and functional capacity of the individual, and (4) supporting the workers professional competence (Ilmarinen 1995). This four-fold approach is most likely ensure the good work ability.

INTERVENTIONS TO SUPPORT HEALTH AND WELL-BEING AT WORK

The number of published interventions to support health, psychological well-being, or the workers' ability to work has grown during the past decade. However, the large variety of interventions and the wide range of outcome measures used, makes it very difficult to demonstrate unambiguous by the results of this research. However, participative approaches to plan, implement and evaluate the outcomes of interventions aiming to improve work or the workers' well-being have been considered to be effective (Elo et al. 1999, Hopsu et al. 1999). On the other hand the aims and outcomes of participative interventions have also been criticized (e.g. Heller et al. 1998). In Finland survey feedback interventions have been used for 20 years to target the interventions and to plan them together with those concerned (Leppänen 1984, Lindström 1990, Hopsu et al. 2000). Still, even more attention should be paid to the process of generating the aims of the interventions (Gherardi 2000).

WORK IN SCHOOL KITCHENS

The work in school kitchens has changed dramatically during the past decade. The food-processing has been rationalized, the work tasks have been reorganized, and productivity demands have increased. According to the personnel, also the mean age of kitchen workers has increased, and their work ability is low. This change has created new demands for occupational skills and professional competence. It is necessary for kitchen workers to be able to evaluate their work tasks carefully and to carry out most of the tasks, avoiding physical and mental overload. In kitchen work, many environmental, organizational and individual factors can lead to impaired work capacity and work disability.

PHASES OF THE DEVELOPMENT PROGRAM

The aim of the program was to improve the work ability and motivation of the kitchen workers.

There was a need to chart the health status of the kitchen workers and the level of the prerequisites for their work ability (e.g. the work characteristics). Table 23.1 shows in chronological order the important "milestones for planning" of the program and its prerequisites. Some principles were decided upon in the basic plan of the program. First, the entire personnel of the school kitchens was to participate in the program. Secondly, the program was to last 1997 to 2000, and thirdly, the results of the program were to be evaluated, data on the program and its progress were gathered continuously. The follow-up study of work ability, its prerequisites and development utilizes the base line measurements. These were repeated at the end of the program in 2000. A large amount of qualitative data was also gathered for evaluation purposes.

Table 23.1 The phases for planning the program

Phases in the planning and realization of the program	The participants
The need to support work ability was recognized	Manager of the kitchen personnel
Decision to conduct the program	Management of the kitchen staff and the Council of the City of the Espoo
Negotiations to implement the program	Representatives of the workers and the management and representatives of the Occupational Health Services (OHS) and Researchers of the Finnish Institute of Occupational Health (FIOH)
Basic plan of the program	Representatives of the workers, of the management and representatives of OHS and FIOH
Setting up a support group for the program	Management of the kitchen staff, FIOH, OHS
Information about the aims of the program to the personnel/ Competition on a name for the program	Management of the kitchen staff, FIOH, OHS
Deciding on a name for the program	Management, on the basis of proposals from the staff
Base line measurements and the report 1997	FIOH planned and carried out the measurements, analyzed the results and wrote the report the fitness tests were conducted by OHS.
Plan of the program	FIOH, manager and the representative of the workers and of OHS

MATERIAL AND METHODS

The study population in the baseline measurements included all the school kitchen workers (n=198) in Espoo, the second largest city in Finland. The mean age of the workers was 44.8 years in 1997, and 98 % of them were women.

The individual factors of well-being, the subjective long-term stress reactions, and the perceived work characteristics, were inquired with a Occupational Stress Questionnaire (OSQ) (Elo et al. 1999). Functional capacity was measured with fitness tests (Liite ry. 1997). The conceptual mastery of the work process was studied with a diagnostic test. The questions focused to the kitchen work was made by the management of kitchen personnel and the researcher of FIOH. The test measures the knowledge of the permanent or potential characteristics of the target system saved in the subject's long-term memory (Leppänen 2001). The items in the diagnostic test of the conceptual mastery of the work in school kitchens included the following themes: economy (7 items), cooking (11 items), nutrition (12 items), kitchen hygiene (8 items), diets (10 items), client service (4 items), hygiene control (15 items), knowledge of the activity environment (18 items).

Needs of development were asked with questionnaire originally used in the development programs in paper industry (Leppänen et al. 1998). There were 18 questions applied to the work of school kitchens, 12 of them concerned with developing cooperation and ways of action. The remain questions were focused on developing tools and environment in kitchen. The Work Ability Index (WAI) (Ilmarinen et al. 1991, Tuomi et al. 1998) was the outcome variable.

The differences in the results of the OSQ and WAI between the age groups were tested with analysis of variances (SAS/STAT user's guide).

RESULTS

Psychological work factors

The assessment of psychological work factors in the entire study group was mostly positive. There were, however, some differences between the age groups in the assessment of psychological work factors. Quantitative workload and boundness to work were assessed as highest by those who were over 45 years of age (Table 23.2). The youngest and the oldest age groups of participants felt that the relations with their workmates were worse than those in the age group of 35-44 year old. The assessment of job control, leadership, the challenge of the work, and role clarity did not differ by age.

Table 23.2 The self-assessed psychological work characteristics and their differences in three age groups of school personnel (M=means, SD=stadard deviation)

Indicators of work characteristic	-34 years M	SD	35-44 years M	SD	45- years M	SD	F value
Job control	3.0	0.5	2.8	0.6	2.9	0.6	0.91 ns
Adequate leadership	2.6	0.9	3.0	0.9	2.8	0.8	0.72 ns
Good relations with colleagues	3.3	0.7	2.9	1.0	3.4	1.0	5.24**
Challenge of work	2.7	0.6	2.6	0.8	2.7	0.8	0.12 ns
Quantitative work load	2.7	0.7	2.9	0.5	3.3	0.7	10.4***
Boundness	2.8	0.7	2.7	0.8	3.1	0.8	4.48*
Role clarity	3.5	0.5	3.5	0.6	3.6	0.5	1.15 ns

***$p<0.001$, **$p<0.01$, *$p<0.05$, # Assessments on scale 1-5, 1=positive, 5=negative.

The older age groups assessed their work as more strenuous both mentally and physically. The oldest group of participants also had more somatic symptoms and assessed their health state as worse than the younger groups of participants (Table 23.3).

Table 23.3 The perceived indicators of well-being by age

Indicators of well-being	-34 years M	SD	35-44 years M	SD	45- years M	SD	F value
Mental strain of work	2.5	1.1	2.8	1.0	3.0	0.9	4.12*
Physical strain of work	3.1	0.9	3.5	0.8	3.6	0.8	5.89**
Mental symptoms	2.2	0.6	2.3	0.5	2.4	0.5	1.62 n.s.
Somatic symptoms	1.7	0.5	1.7	0.5	2.0	0.6	3.91 *
Health state compared with others of the same age	1.9	0.6	2.0	0.9	2.4	0.8	4.76**

***$p<0.001$, **$p<0.01$, *$p<0.05$, # Assessments on scale 1-5, 1=positive, 5=negative.

Need to develop work processes in school kitchens

More than 40% of the participants felt that it would be important to improve their physical fitness and the co-operation between the workers in one kitchen. 39% of the respondents felt that it would be useful to improve their professional skills. The oldest group stressed the importance of improving these issues even more than the two younger groups (Figure 23.1).

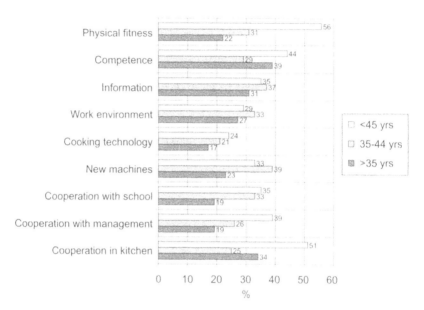

Figure 23.1 Needs to develop the work process in school kitchen in different age groups

Professional competence

The participants felt that their professional skills ought to be improved. The results of the diagnostic test also indicated that the mastery of the work process should be developed further. The average level of the right answers varied from 52% to 73% between the eight items of knowledge in the kitchen work.

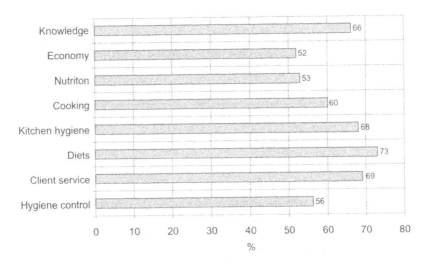

Figure 23.2 The diagnostic test of knowledge of the main items in kitchen work
(the means of the percent of the right answers)

Work Ability Index (WAI)

The current work ability of over 45-year old kitchen workers had deteriorated 22% of
them felt that their current work ability was quite low (Table 23.4). On the other hand,
the workers who were under 45 years felt that their work ability was good compared
to life-time best.

Table 23.4 The distribution (%) of the current work ability compared to the best
of lifetime (0=completely unable to work; 10=work ability at its best)
in the different age groups

Current work ability	18-34 yrs (n=30)	35-44 yrs (n=53)	>45 yrs (n=107)
9-10	64	38	16
7-8	32	50	61
<6	11	12	23

Most of the workers in the younger age groups had an exellent or good work
ability index, but 60% of the kitchen workers over 45 years old had an average or
poor work ability index (Table 23.5).

Table 23.5 The Work Ability Index (WAI) categories in the different age groups

WAI	18-34 yrs (n=30)	35-44 yrs (n=53)	>45 yrs (n=107)
Excellent	28	13	4
Good	47	58	37
Moderate	22	25	47
Poor	3	4	13

THE PROCESS

It is necessary to make such an exhaustive work before the program is ready. Without baseline measurements and careful planning the development process does not go on. And we also know when the management and the workers planned (with the help of researchers) the program together they both also commit themselves to carry out the development process. So the intervention program was based on the baseline results. The results showed that the physical strain of the work process should be decreased, the professional qualifications of the workers should be improved, the co-operation in the kitchens should be improved, and one of the most important tools in kitchen work, the worker's body, should be taken better care off. Futhermore, 60% of the over 45-years old workers were in the poor or moderate category of the Work Ability Index, which means that without additional support, they might have great difficulties in coping with their work in the near future. In this process, the development of the work included also the training of the management. The intervention was based on the participatory approach and it included arranging the work new ways and improvements in the work methods as well as team work. The following program was undertaken to reach these objectives (Figure 23.3). The intervention was started at the beginning of 1998, and evaluated in 2000.

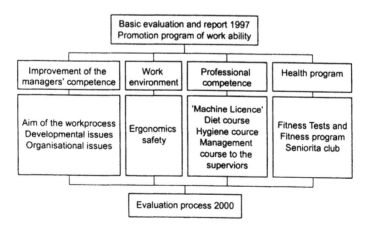

Figure 23.3 The intervention process

REFERENCES

1. Elo, A-L., Leppänen, A., Lindström, K. and Ropponen, T., 1992, OSQ; Occupational Stress Questionnaire:User's Instructions. Institute of Occupational Health, Helsinki, Finland, pp. 43 + appendices.
2. Elo, A-L., Leppänen, A., 1999, Efforts of health promotion teams to improve the psychosocial work environment. *J. Occup. Health Psychol.* 4: 2, pp.87-94.
3. Gherardi, S., 2000, Where learning is: Metaphors and situated learning in a planning group. *Hum. Rel.* 53: 8, pp.1057-1080.
4. Hopsu, L., Louhevaara, V. and Korhonen, O., Effects of ergonomic and exercise intervention on stress and strain in professional cleaning. FinnAge_Respect for the aging. Action programme to promote health, work ability and well-being of aging workers in 1990-96. *People and Work. Research reports* 26, pp.138-146.
5. Leppänen, A., 1984, Reduction of stress by personnel at institutions for child care and for the mentally handicapped. In *Stress and tension control 2*, edited by F.J. McGuigan, W.E. Sime J. and Mac Donald Wallace, (New York: Plenum), pp.285-294.
6. Leppänen, A., 2001, Improving the mastery of work and the development of the work process in paper production. *Relations indutrielles- Industrial relations 56:* 3, pp.579-609.
7. Leppänen, A., Elo, A-L., Ilmarinen, J. and Alanko, O., 2001, Parempaa työkykyä etsimässä. (In search of means to improve the employees' ability to work) In *Finnish with an English Abstract. Work and people. Research report 19*, pp.94.
8. Lindström, K., 1995, Finnish research in organizational development and job redesign. In *Job stress interventions*, edited by L.R. Murphy, J.J. Hurrell jr., S.L. Sauter and C.P. keita, APA, Washington D.C., pp. 283-294.
9. Liite RY, 1997, Kuntotestauksen Peruateet. Liikuntalaaketieteen Ja Testaustoiminnan Edisteeámisyhdistys, Helsinki.
10. Ilmarinen, J., 1995, Aging and Work: The Role of Ergonomics for Maintaining Work Ability During Aging. In *Advances in Industrial Ergonomics and Safety VII*, edited by A.C. Bittner Jr. and P.C. Champney (UK: Taylor & Francis), pp.3-17.
11. Rothstein, L.R., 1992, The Case of the Temperamental Talent. *Harvard Business Rev. Nov-Dec.*, pp.16-26.
12. SAS Institute Inc., 1986, SAS/STAT User's Guide Version 6, 4th edition, Vol. 2, pp.1633-1640.
13. Tuomi, K., Ilmarinen, J., Jahkola, A., Katajarinne, L. and Tulkki A., 1994, Work Ability Index, Institute of Occupational Health, Helsinki, p.18.

PART V
Support Systems for Elderly Workers

24. Developing a New Work System for Aging Workers

Mitsuyuki Kawakami

Department of Amenity Ergonomics, Tokyo Metropolitan Institute of Technology, Tokyo, Japan

ABSTRACT
The ratio of aging to young people in the world is increasing due to better medical care and changing lifestyles. However, this has also resulted in a considerable reduction in labor productivity, as aging people remain active in the workforce for longer periods. Japan is rapidly becoming an aged society at an unprecedented rate that is clearly unlike the experience in other countries. In the Japanese labor environment important negative factors such as the rapid fluctuation in the exchange rates and the aging of our labor force exist creating difficulties for production related industries. Some companies have resonded to fluctuating exchange rates by introducing industrial robots to increase automation while others have increased overseas production in an effort to trim labor costs. Overseas production, however, has results in rising production costs and blue-collar unemployment in Japan. From the production standpoint, manpoauthorr is still the core of man-machine systems. The aging of our labor force is not amenable to administrative interventions. Currently, university research institutes, the government, and enterprises are grappling with solutions to this problem.
As background, this report presents a proposal to developing elements of a new work system and the concept for aging workers.

Keywords: *Aging workers, Productivity, Workload, Job enlargement, Job enrichment, Information technology, New work system*

INTRODUCTION
In the Japanese labor environment important negative factors such as the rapid fluctuation in the exchange rates and the aging of our labor force exist creating difficulties for production related industries.

Some companies have responded to fluctuating exchange rates by introducing industrial robots to increase automation while others have increased overseas production in an effort to trim labor costs. Overseas production, however, has resulted in rising production costs and blue-collar unemployment in Japan. From the production standpoint, manpower is still the core of man-machine systems (e.g., the TOYOTA Motor Co. manpower system).

The aging of our labor force is not amenable to administrative interventions. Japanese society is aging at an unprecedented rate of four times that reported in Europe and North America. The Population Research Institute of the Welfare Ministry estimates that people over 65 years old will form 25% of Japan's population in 2020 (Nagamachi, et al., 1985)[1]. This means that our country will have the largest ratio of aging to young citizens in the world, resulting in a considerable decline in labor productivity if we continue using current manufacturing techniques. To address this problem the government research laboratories and other institutes are jointly cooperating in researching matters related to the employment of an older workforce.

Although many investigators, including the present authors, have researched the means for maintaining productivity among workers as they age (Miyashiro, 1985. Miyashiro, and Yokomizo, 1987. Shibata, et al., 1993. Kawakami, et al., 1986. Nagamachi, et al., 1983)[2-6], few studies have focused on the elements involved in job area redesign for older workers. This report presents a proposal for the redesign of job area elements based on an experimental model that author have developed.

CONCEPT OF JOB DESIGN FOR AGING WORKER

The model used to conduct this research is shown in Figure 24.1. The conceptual framework of the model is based on an ongoing series of research projects conducted by the author.

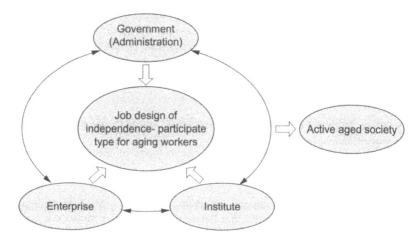

Figure 24.1 The concept of job design for aging workers

CHARACTERISTICS OF AGING PERSONS

A number of general characteristics seen among aging workers were defined as follows (Nagamachi, et al., 1985; Nagamachi, Kawakami, Une, & Tange, 1983)[1,6]: (1) Vital functions such as sight and hearing decline rapidly in proportion to age; (2) a decline in the muscles of legs followed by those in the waist, shoulders, arms, and then the hands respectively; (3) the muscles and functions for those who train or exercise decline more slowly than for those who don't exercise. The decline is still noticeable, however; (4) older workers' technical skills and judgment, accumulated over years of experience, are superior to those of younger worker; (5) as workers get older, their individual differences grow; and (6) it becomes increasingly more difficult for older workers to adapt to a predetermined production line speed in a workshop.

The above points illustrate both positive and negative factors that must be considered in the design of work systems for older workers. Our concern was to design a work place that was compatible with both the positive and negative factors discussed above and yet remain competitive in our production costs. To do this author used an approach based on a Worker Oriented Job Design.

BASIC IDEA OF WORK SYSTEM DESIGN FOR AGING WORKERS

The basic ideas of the developing a new work system for aging workers are as follow. A premise of the system design plan to solve the observed problems was created using components of the Y theory (McGregor, 1960)[7]; using the classification of needs, order of importance, proposed by Maslow, P.H. (Maslow, 1954)[8]; using the principles of motion economy as related to the work place by Ralph M. Barnes (Barnes, 1937)[9] and using previous research on systems design for the aged workers by authors (Kawakami, Inoue, Ohkubo, & Ueno, 1997)[10].

THE INFLUENCE OF AGE DISPARITY

Generally in various motion elements of "get and place" activity are a suitable operation for explaining the difference of motion characteristics according to age because special skill is not needed (Get: the worker's hand reaches for object and grasps it, Place: the worker's hand moves the object to working point).

As can be seen in Figure 2, changes of the motion velocity waves (motion velocity waves are the change of hand speed in the "get" task or "place" task) of young workers and older workers were observed over a "get and place" operation by 3D motion analyzer. Human motion velocity is generally composed of acceleration ® constant-speed ® deceleration (Mandel, 1961)[11]. This figure shows that motion velocity waves are divided among Phase I (acceleration), Phase II (constant speed) and Phase III (deceleration). It appears from Figure 24.2 that the younger worker's operation is composed of Phase I-II-III pattern (I.e., acceleration ® constant-speed ® deceleration), but the aged worker's Phase I-II-III operations were all delayed when compared to the results from the younger cohort. Moreover, the transition of motion velocity for the older workers, from acceleration to deceleration occurred at a slower terminal velocity than that seen among the young workers, a finding that author believe an age related due to the decrease of human functions (especially muscle function) resulting from the aging process (M. Kawakami et al., 2000, 2001)[12,13].

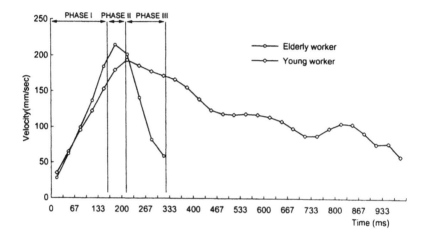

Figure 24.2 Motion velocity wave of get and place activity

As can be seen in Figure 24.3 change of the motion locus of young workers and aging workers were observed over a "get" and "Place" task by 3D motion analyzer. This figure shows that the younger worker's operation locus are a smooth and rhythmical, but the aging worker's operation locus are a rough and non rhythmical.

As can be seen in Figure 24.4 change of the eye fixation point on locus of young workers and aging workers were observed over a fasten lead and tighten up screws task was observed by the authors using an eye camera. This figure shows that the younger worker's eye fixation point and locus are short time and gathering, but the aging worker's there are long time and widely scattered.

Figure 24.3 Comparison between a motion locus of aging and young worker

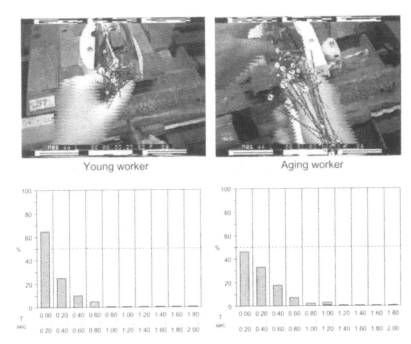

Figure 24.4 Comparison between an eye fixation point on locus of aging and young worker

CASE STUDY OF DEVELOPING A NEW WORK SYSTEM FOR AGING WORKERS

The method was to abstract and analyze the problems of the Production Flow Line System (PFLS) that has been operating using a four-person assembly and two-person inspection process, proposing a new System and verifying the efficiency of the system through the experiment discussed below[14].

The methods of experimentation for this investigation involved (1) a Time study (Motion time and Cycle Time of Product) and (2) a Work Load Examination (CFF: Critical Flicker Fusion Frequency and Subjective Symptoms of Fatigue). The subjects of this study worked on an assembly line producing a vaporizer for oil fan heaters in a plant of a well-known electrical appliance manufacturer. The average age of workers (women) was 47.3 years.

This research identified the following problems. (1) Productivity: the average ratio of achievement was 78% as compared with the production goal quantity set for the day. (2) The efficiency of line balance: 64.9%. (3) Workload: noticeable fatigue from dullness caused by repetitive motion activity was recorded via means of the current system (PFLS) is shown, Figure 24.5.

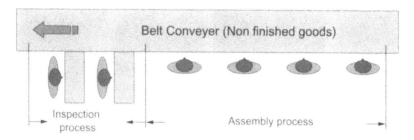

Figure 24.5 The Style of Flow Line System (Current System)

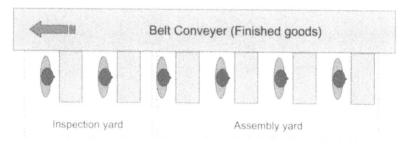

Figure 24.6 The Style of Job Enlargement System (New System)

The proposed "New System" is expected to result in (1) improvement in productivity, (2) decrease in the workload burden (particularly fatigue from dullness), (3) respect for workers experientially judgment and decisions.

The system eliminates the PFLS and provides for job enlargement where workers can use 100% of their abilities. To do this "Plan" and "Check" functions are added to each operation. In this way each worker is asked to perform the function of Plan-Do-Check action. Due to the age characteristics of aging workers a function of "Careless Prevention and Out Put Self-Control" was conceived for this system to allow workers to evaluate their own performance. The new system is shown Figure 24.6. This system has been enlarged with the four assembly process of flow line system. As a result of the New System, one worker performs the tasks that four people had in the previous system. A comparison of production results for both systems is shown Figures 24.7 and 24.8.

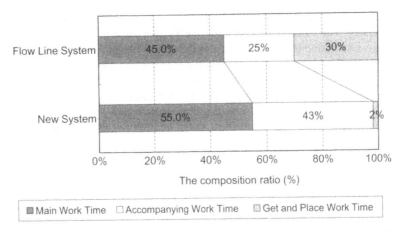

Figure 24.7 The change of composition ratio on each time taken to assemble

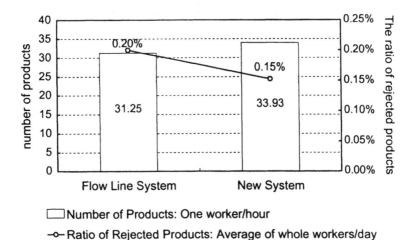

Figure 24.8 The number of products and the ratio of rejected products

The total per product assembly time for the PFLS was 39 seconds and for the New System 31 seconds. The ratio of improvement was 21%. Figure 24.7 shows that the New System allowed for a 28% decrease in the ratio of Get and Place Time to Movement Time. Namely as a result of work enlargement, the decrease of Get and Place Activity reduces the Total Assembly Time.

A comparison of the final production output numbers is shown in Figure 24.8. It can be seen that the New System showed improved output number and reduced number of rejected products while making workers perform the Plan and Check functions. The whole line balance was improved to 92.6%.

In Figure 24.9, the comparison of the change of CFF in the PFLS and New System are indicated. The workers in the Flow Line System perform their repetitive jobs creating the condition of dullness, whereas the workers using the New System show acceleration of pace. Author believes this is due to the New System's propensity for respecting worker independence and judgment resulting in self-control. As can be seen in Figure 24.10, the results of the Subjective Symptom of Fatigue examination showed a decrease of 20% on the assembly line using the New System compared with the data from the PFLS. The results show author improved workplace design for aged workers reduces fatigue.

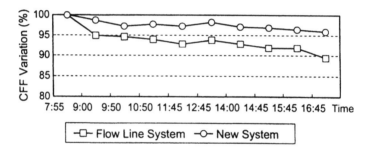

Figure 24.9 The comparison of CFF in a day on the Flow Line System and New System

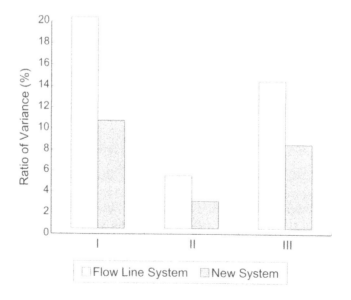

Figure 24.10 The result of survey of subjective symptom

CONCLUSION

From the report presented above, the following points are considered important:

(1) As a developing a new work system for aging workers, it is necessary to do construction the concept for aging workers.
(2) The workers can reach 'self-recognition' and the 'self-actualization' level seen in Maslow's five classifications need theory.
(3) The most important concept is that this system allows for harmony the productivity of a work system according to the needs demonstrated by author's aging work-system in a human yet productive way.

This is necessary not only to improve productivity in manufacturing but also to make the lives of our aging workers more productive by allowing companies continue to tap the valuable work resources that these aging workers possess.

REFERENCES

1. Nagamachi, M., Kawakami, M., Kumashiro, M., Une, S. and Tange, S., 1985, Textbook Job Redesign. In *The Association of Employment Development for Senior Citizens*, Japan, pp. 7-119.
2. Miyashiro, N., 1985, Control Process in Upper Extremity Movement from the Accelerating Viewpoint. In *Journal of Japan Industrial Management Association*, **36** (3), pp. 202-207.
3. Miyashiro, N. and Yokomizo, Y., 1987, Relationship between the Visual Feedback Mechanism and the Task Difficulty in Simultaneous Motion. In *Journal of Japan Industrial Management Association*, **38** (1), pp. 34-35.
4. Shibata, S., Ooba, K. and Inooka, H., 1993, Experimental study on human upper link point-to-point movements which require positioning accuracy. In *The Japanese Journal of Ergonomics*, **29** (5), p. 281.
5. Kawakami, M. and Ueno, T., 1986, A Comparison on the Number of Work Station Divided between Two Different Work Systems. In *Journal of Japan Industrial Management Association*, **37** (5), pp. 295-302.
6. Nagamachi, M., Kawakami, M., Une, S. and Tange, S., 1983, Job redesign encyclopedia for senior citizens. In *The Association of Employment Development for Senior Citizens*, Japan, pp. 1-328.
7. McGregor, D., 1960, The human side of enterprise. In *McGraw-Hill Inc.*, New York.
8. Maslow, P.H., 1954, Motivation and personality. In *Harper and Row*, New York.
9. Ralph M. Barnes, 1937, Motion and Time Study 4th Edition. In *John Wiley&Sons Inc.* Newyork, pp. 247-301.
10. Kawakami, M., Inoue, F., Ohkubo, T., Ueno, T., 1997, Job Redesign Need for aged workers. In *Proceedings of the 13th Triennial Congress of the International Ergonomics Association*, vol.5, pp. 448-450.
11. Mandel, M.E., 1961, Motion and Time Study, Prentice-Hall, Inc. New York, pp. 505-508.

12. Kawakami, M., Inoue, F., Ohkubo, T. and Ueno, T., 2000, Evaluation elements of the work area in terms of job redesign for older workers. In *International Journal of Industrial Ergonomics*, **25**, pp. 525-533.

13. Kawakami, M., *et al.*, 2001, National project: A study of developing for Active aging model in manufacturing industry. In *The Association of Employment Development for Senior Citizens*, Tokyo, Japan.

14. Kawakami, M., Inoue, F. and Kumashiro, M., 1999, Design of a work system considering the needs of aged workers. In *Experimental Aging Research*, 25, pp. 477-483.

25. The Theory and Practice of Work Re-design in Small and Medium-sized Manufacturing Enterprises in an Aging Society

Koki Mikami

*Department of Information Network Engineering,
Hokkaido Institute of Technology, Hokkaido, Japan*

ABSTRACT

Japanese society is now rapidly aging because of the current drop in the birth rate. The younger working population is decreasing, and the issue of how to utilize the ability and experience of middle-aged workers is extremely important with respect to the continuance of companies, the assurance of the individuals' living and national pension finance. The decline of physiological functions caused by aging is undeniable. "KAIZEN" is one of the measures which make it possible to reduce workload, which is an obstacle in the way of elderly peoples' continued employment, to have them keep working in good health and comfort, and to maintain a highly productive workplace. In this report are described practical cases of workplace re-design in small-scale manufacturing using Ergoma Approach and improvement tools which the author has been using, and also the importance of "KAIZEN" know-how database construction and its horizontal development in an aging society.

Keywords: *Aging, Healthy companies, KAIZEN, Know-how database*

THE LABOR FORCE PROBLEM OF "SUPER-ADVANCED AGE AND FEWER CHILDREN" SOCIETY

In Japan, after the collapse of the "bubble" economy, companies have reduced their personnel as a restructuring and rationalizing measure, and in many cases people of middle and advanced age have become the objects. Considering the labor force composition of our country, however, a decrease in the younger labor force will be inevitable in several years, and securing younger laborers is now becoming a serious problem in smaller companies with many dirty, dangerous and demanding jobs. As one more important aspect, we should understand that a decrease in quantity means quality problems, which are thought to influence companies' assurance of a satisfactory work force both in quantity and in quality in the near future.

INDISPENSABLE CREATION OF HEALTHY COMPANIES

To make Japan's coming "super-advanced age and fewer children" society into a vigorous society, construction of an industrial labor system which makes the following three points possible will be an important issue.

1) In society, increase of tax revenues from individuals and companies, reduction of social security expenses, and sound national finance.
2) For the individual, that people with the will and ability can continue working in comfort as long as they wish.
3) In the company, with fewer younger workers, much more improvement in productivity by aging workers with rich experience and sufficient ability both in quantity and quality.

As the management of a company examines continued employment of aged workers, the following points are raised: 1) Re-examination of the seniority-based system and the employment managements such as wage structure, 2) establishment of educational training and human resources development systems relating to adaptation to work and support for self-education, 3) positive support for health care by the company, 4) re-examination of the re-employment system up to the public pension provision age, and 5) creation of workplaces friendly to workers of advanced age. Moreover, the administration is also implementing many support measures toward job security of people of middle and advanced age. However, the most important premise in a "super-advanced age and fewer children" society is that the company itself must be secure, with a strong constitution. Considering the quantity or quality of the younger labor force in Japan's aging society, it is essential to aim for the coexistence of productivity and humanity and create a highly productive, healthy company in a true sense, where it is possible to make up for the lowering of middle-aged workers' fluid ability, have them demonstrate their accumulated skills and continue employing them.

SPECIFIC PROCEDURE OF WORK RE-DESIGN

1) Ergoma Approach

KUMASHIRO [1] raises rationalization of working conditions and work environment (including work re-design and support apparatus) as one of the four keywords in the basic strategy of companies dealing actively with the aged society. "KAIZEN" is one of the measures which make it possible to reduce workload, an obstacle in the way of older people's continued working, to have them keep working in good health and comfort, and to realize a workplace of high productivity. The author et al. [2-3] have used Ergoma Approach aiming for the fusion of productivity and humanity in our research about work re-design based on "KAIZEN". Ergoma is a coined compound term composed of "ergo" from ergonomics and "ma" from management. This approach is for improvement of workplaces from the viewpoint of industrial psychology, such as the awareness of duty and satisfaction level of workers, as well as the viewpoint of ergonomics such as work burden, work postures, work environment and the conventional viewpoint of industrial engineering (IE).

Fig. 25.1 shows the procedure for Ergoma Approach.

Step1 ;	**Aim**
	Pick out long-or-short-term problems for the company to solve

Step 2 ;	**Whole-company-tackling**
	Organize an improvement project team consisting of employers, managers and workers

Step 3 ;	**Preparatory investigation**
	Hold a hearing to listen to workers' opinion, and conduct a preparatory observation in the workplace

Step 4 ;	**Discovery of problem worksites and items**
	Seize target worksites or tasks awaiting solution, and clear up the causes

Step 5 ;   **Analysis of the present conditions**

Do research in the target worksites taking into account the indication matters of Step 6 from the viewpoints of IE, Ergonomics and industrial psychology. Choose approaches appropriate for the target sites. The following are the items the author et al. usually use.

[IE aspect] Operation ratio, Analysis of flow, Analysis of rink, Analysis of Layout etc.

[Ergonomics aspect] Psycho-physiological function, Fatigue, Analysis of work posture etc.

[Industrial psychological aspect] A survey of job consciousness, health and working conditions, Workers' opinion about improvement of labor environment

Step 6 ;   **Indication matters**

Classify the results of the present conditions into the following items and indicate the direction of "KAIZEN"

The influence of work on humans & The influence of humans on work

1) Unsafe operations 2) Workload 3) Unsafe conditions 4) Work content 5) Uncomfortable working postures 6) Health	1) Labor productivity 2) Job satisfaction	1) Unsafe actions 2) Job consciousness 3) Degree of concern for work 4) 5 S's 5) Management conditions 6) Workers' background

Step 7 ;   **Examination of "KAIZEN" plans**

The "KAIZEN" project team discovers the true cause from step 6, makes "KAIZEN" plans, and examines them. The use of work improvement support tools such as Work Posture Burden Evaluation System, Support Information System for creating new ideas and Virtual Simulation enables effective improvement.

Step 8 ;	**Practice of "KAIZEN"**
	Incorporate the improvement plans. Ensure that all the workers practice the improved work.

Step 9 ;	**After-improvement evaluation**
	Measure the after-improvement effects. Especially hearing of workers is important for next "KAIZEN".

Figure 25.1 Ergoma approach

When detailed present-data-analysis technique is impossible in the company which is going to practice "KAIZEN", we can easily point out problems by observing the workplace from the viewpoint of each item of the pointsshown in Step 6. It is important to explore the true causes one by one with Step 6 as a key, to work out measures for improvement using the wisdom of the group members, to decide improvement time limits and who should be in charge, and then to solve the problems.

2) Support tools for "KAIZEN"

Practice is important in "KAIZEN". Step 7 of Ergoma Approach is where the true cause is pursued from the points shown, and plans are made for improvement practice. However, there are many cases where "KAIZEN" fails to be practiced because the people do not know what ideas there are, what is an effective way to begin with, or whether the support apparatus in the improvement plans are effective or not, or just because the workers themselves dislike changes in the working procedure. This is common particularly in small and medium-sized enterprises. In order to solve these problems, the author et al. [4-7] have developed and used three support tools for "KAIZEN" so far.

One of the tools is a work posture evaluation system which enables the selection and evaluation of a preferential improvement plan using work postures as a means [4-6]. The second tool is an advanced evaluation system of investment effects, which evaluate the effectiveness of an improvement plan using virtual simulation [6-7]. The third is an improvement case information system using the Web, where "KAIZEN" cases are databased [8]. It enables the creation of ideas and a check on effectiveness. The KAIZEN cases are from the joint research conducted from 1986 to 1997 by the Association of Employment for Senior Citizens, a body founded in 1978 by the Ministry of Labor (AESC) and companies, and content retrieval is possible by linking to the association.

3) Practice of "KAIZEN"

Some of the examples of "KAIZEN" practice in connection with work re-design in a small-scale manufacturing enterprise are described here.

PRACTICE CASE 1.

Job re-design for continued employment was performed in a concrete-product manufacturer whose workers were advancing in age [6-7]. As a result of Ergoma Approach, the following nine tasks were the targets of support apparatus instruction from the viewpoints of workload reduction and improvement in productivity.

1) Washing of concrete injecting machines, 2) Injecting of ready-mixed concrete, 3) Welding of reinforcing bars, 4) In-factory cleaning work, 5) Transporting of mold release pumps, 6) Transporting of tools, 7) Covering and removing of sand and gravel tents, 8) Loading and unloading of products by truck/crane, 9) Adjusting the fork

width of forklifts.

(1) Use of the work posture burden evaluation system

Fig. 25.2 shows the work posture burden evaluation indices of these tasks by the work posture burden evaluation system.

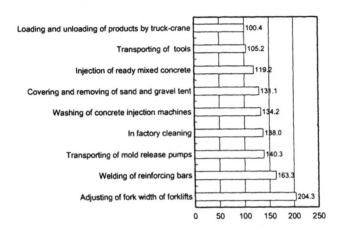

Figure 25.2 The work posture burden evaluation indices

(2) Employee-participation-type decision on support apparatus plans

Support apparatus plans for these tasks were made with participation of the employees. For example, "Washing of concrete injecting machines" was a task in which workers wash the machine using a deck brush and a hose after ready-mixed concrete injection. The problem there was that washing took time, with much load on the arms and lower back of the workers, and it was an unsafe task with bad scaffolding. As support apparatus, plans for a high-pressure water machine and a work stand were made.

(3) Advance evaluation of improvement plans using the advance evaluation system for investment effects

Using the system, "present-work" simulation and "work-after-improvement" simulation were conducted, and then the improvement plans were evaluated. Fig. 25.3 shows each simulation screen. It was demonstrated that the total consumed Kcal was 9.42 in the present-work simulation and 1.21 in the work-after-improvement simulation. These results confirm an 87% decrease in the total consumed Kcal. Moreover, it became evident that RMR could be decreased by two stages from heavy work of 5.77 to light work of 1.45

From the results of the above quantitative evaluation indices of workload and visual comparison, we concluded that the investment effects of the support apparatus would be very large in the workplace, too.

Present-work    Work-after-improvement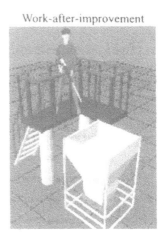

Figure 25.3 Each simulation screen

(4) Introduction of support apparatus

Washing of the concrete injection machine after introduction of the support apparatus is shown in Fig. 25.4.

There was decrease verified in the effect measurement of subjective feelings of fatigue, heartbeat and work posture load evaluation indices. In the other tasks, too, reduction of workload and shortening of working hours were confirmed, which verified their effectiveness with regard to improvement in productivity.

Figure 25.4 Washing after introduction of the support apparatus

(5) The results of "KAIZEN"

In this company, "a standard work combination pitch diagram" in connection with productivity, health maintenance, etc. and "a work procedure manual" in connection with safety, quality, etc. were made in addition to introduction of the support apparatus. As a result of "KAIZEN", we were able to create a workplace of high productivity where older people could continue working in good health and safety.

(6) Usefulness of the "improvement case information system"

The system developed jointly with the Association of Employment Development for Senior Citizens is thought to be useful in 1) collection of ideas from many fields regardless of type of industry, 2) effectiveness conformation of ideas, and 3) creation of more advanced ideas. After completion of the system [8], we input the "washing work" into the retrieval screen in order to create better ideas for washing of the ready-mixed concrete injecting machine. The retrieval result is shown in Fig. 25.5. There was one successful improvement cases by Mishima Kosan Inc. (1994). The case showed that extension of the washing space and introduction of a flexible slide shutter as well as introduction of a high-pressure water cleansing machine enabled them to achieve the reduction of uncomfortable work postures. Fig. 25.6 shows comparison of the picture from the retrieval screen and the above-mentioned work. There is a similarity between the two improvement plans, which shows that a user can obtained ideas from different industries and confirm their effectiveness in advance.

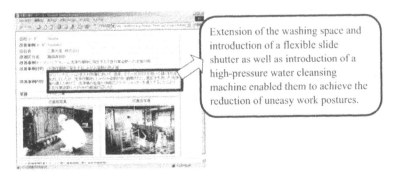

Figure 25.5 The retrieval result

Figure 25.6 The comparison of the picture from the retrieval screen and the washing of the concrete injection machine

PRACTICE CASE 2.

Work re-design was performed for continued employment at a wire-fence manufacturer [9].

(1) The results of Ergoma Approach analysis

A total of 45 problems were discovered by Ergoma Approach analysis in the target company. The details of the analysis were given to the employees. Then the problems were tackled in cooperation with them (employee participation) by working out measures and appointing a person in charge and a time limit for improvement.

(2) Eighteen improvement items were executed.
1) Development of a twin edge processor
2) Preparation of an edge processor assembly (development) manual
3) Preparation of computer-animated operation manuals
4) Confirmation of member stock
5) Confirmation of member storage places
6) Making of carts for stacking and carrying finished products
7) Making of turntables to stack finished products on
8) Making of carts to carry members
9) Preparation of criteria for the management of the combination of "jirasu" and "hera" attachments
10) Preparation of an adjustment manual for edge processors
11) Making of carts for carrying tools
12) Preparation of a workbench combination list for production
13) Installation of lighting equipment on machine No. 1
14) Installation of the net-making machine's hydraulic power unit for cutting lines
15) Preparation of a management manual for the hydraulic cutting unit
16) Installation of handrails on the stairs leading to the rest room
17) Height adjustment of lighting equipment
18) Installation of a factory entrance door

(3) Preparation of computer animated operation manuals

In preparing the manuals 3), we obtained ideas through the "improvement case information system". The work of the first assembly process in this company was mainly conducted manually by experienced female part-time workers working in groups. Although theses workers could assemble almost all the types of products, their work procedures were not necessarily the same and depended on the skills of each individual worker. Also, part-time workers at this company finished working and handed over their work to the second process workers at 3:00 p.m. The second process workers, however, did not have sufficient knowledge about the operation of the first process and could not process all the types of products. Therefore, we hoped to create ideas about transmitting the company's product processing expertise, speeding up the operations and teaching more diverse skills to the full-time workers and workers who joined the company halfway through the year. We input the word "manual" in the retrieval screen. Fig. 25.7 shows the results. There was a successful improvement case by Orio Ironworks Inc. (1995). The case showed that for the purpose of developing a system which helps to hand down smoothly the skills of middle-aged and older workers and skilled workers, they prepared a digital camera manual, which enabled them to reduce the trouble of handing down their skills by word of mouth as they had done before, and to make up for individual differences in writing ability. Based on this, we prepared operation manual CDs using computer animation to educate and train workers in this improvement. In making the manual, computer animation was used in order to make the manual easier to understand than a text-only manual and also to raise the employees' awareness of work improvement by awakening them to the arrival of the information age and the most up-to-date technologies. In this research, the following operation manuals were prepared with computer animation. 1) an operation manual for gabions, 2) an operation manual for double-panel baskets, 3) an operation manual for side frames of items with atypical shapes.

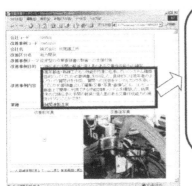

For the purpose of developing a system which helps to hand down smoothly the skills of middle and old age workers and skilled workers they prepared a digital camera manual, which enabled them to reduced the trouble of handing down their skills by word of mouth as they had done before and to make up for individuals' difference in writing ability.

Figure 25.7 A successful improvement cases by Orio Ironworks Inc.

(4) The results of "KAIZEN"

In this company, through the employees' requests many types of support apparatus were made on an experimental basis and the requested improvement was actualized, which reduced their workload, improved productivity and ensured safety. That also led to the creation of a favorable workplace environment for middle-aged and older workers. Further, the foundation for improving the company's development techniques, transmitting their product processing expertise and teaching/training concerning equipment improvement was established by preparing assembly and adjustment manuals themselves for the support apparatus developed by the company, as well as a computer animated operation manual. In addition, the awareness of work improvement was raised and the relative occupational abilities were improved mainly among the middle-aged and older workers.

THE SIGNIFICANCE OF HORIZONTAL DEVELOPMENT OF THE "KAIZEN" KNOW-HOW DATABASE

It is true that work improvement, mostly with regard to improvements in productivity, is carried out within each company and is not often available to people outside. Moreover, although work improvement is easy for companies with improvement know-how and staff, many small and medium-sized companies do not know how to move forward with it concretely. However, under the circumstances where it is indispensable to strengthen the measures to create workplaces that deal actively with the aged society. for older people to continue working there as well as to improve productivity, it is necessary to provide an efficient know-how database of "KAIZEN" so that anyone can access the accumulated know-how easily. The author currently is heading a project that is part of the Ministry of Health, Labor and Welfare's contribution to the national Millennium Project, "Research into Construction of a Work Improvement Support System using the Web". The concept of the system is shown in Fig. 25.8. Development of this system will make it possible for anyone to share the know-how of how to utilize older people anytime, anywhere, and for "KAIZEN" to be incorporated horizontally in each individual company. Creating such healthy companies is what is important in an aged society.

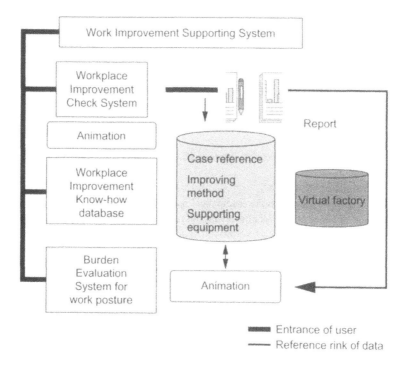

Figure 25.8 Concept of a Work Improvement Support System using the Web

CONCLUDING REMARKS

In this report I have stated the "KAIZEN" technique by developed the author et al., some cases of our tools and work re-design and the outlines of the project we have been working on. Each company has much room to increase its efficiency, and has the potential to increase its added value by integrating productivity with human factors. "KAIZEN" is said to be eternal and infinite, which means that one improvement will produce another problem and that this repeats itself. The newly produced problem certainly is not as serious as the previous one, which leads to progress. There is no question that "KAIZEN" makes the company grow and helps the employees to do their job better. Healthy companies and employees lead to their economic security. When the work improvement support system using the Web which we are developing now is put to practical use, and developed horizontally, and when actualized know-how and knowledge about utilization of older people are accumulated there, there will be manypossibilities for new creation. In the coming "super-advanced age and fewer children" society, further and continuous "KAIZEN" by the company, the individual and the state will be indispensable.

REFERENCES

1. Kumashiro, M., 2000, Ergonomics strategies and actions for achieving productive use of an aging work-force. In *Ergonomics*, **43**(7), pp. 1007-1018.
2. Mikami, K. and Kumashiro, M. *et al.*, 1997, A scientific Approach to Work Improvement (I) -From the viewpoint of Ergoma Approach and Virtual Simulation. In *The 14th International Conference on Production research*, Osaka, Japan, pp. 1152-1155.
3. Mikami, K. and Kumashiro, M. *et al.*, 1997, A scientific Approach to Work Improvement (II) -A case Study with Ergoma Approach and Virtual Simulation. In *The 14th International Conference on Production research*, Osaka, Japan, pp. 1156-1159.
4. Mikami, K., Kumashiro, M. and Izumi, S., 1996, One Approach to Quantitative Evaluation of the Skeletal and Muscular Loading Resulting from Work Postures, and Ergonomic Improvements to Reduce Loading: ADVANCES IN OCCUPATIONAL ERGONOMICS AND SAFETY. In *Proceedings of the XI Annual International Occupational Ergonomics and Safety Conference*, pp. 366-371.
5. Shibuya, M., Mikami, K. and Kumashiro, M. *et al.*, 1998, Development of an Evaluation System for the work posture Burden and Introduction of the System to Worksite. In *Global Ergonomics, ELSEVIER*, pp. 357-360.
6. Mikami, K., Shibuya, M. and Kumashiro, M. *et al.*, 1998, A supporting system for work improvement to create a high-productivity workplace -A System for evaluation the burden of work postures and virtual simulation. In *Global Ergonomics, ELSEVIER*, pp. 497-500.

7.   Mikami, K., Shibuya, M. and Kumashiro, M. *et al.*, 1998, Improvement in Work Environment by A Virtual System From The Viewpoint of Restoration of Human-Nature. In *Ergonomics practice and Its theory: The 5th Pan-Pacific Conference on Occupational Ergonomics*, pp. 188-191.
8.   Kumashiro, M., Mikami, K. and Hasegawa, T. *et al.*, 2001, A Database of Success Stories on "KAIZEN" Activities for Elderly workers. In *Proceedings of 2000 Spring Conference of ESK and International Symposium on Ergonomics*, pp. 253-257.
9.   Mikami, K. Shibuya, M. and Kumashiro, M. *et al.*, 1999, A practical research on workplace friendly to workers-A wire-fence Manufacturing Workplace. In *Proceeding of the 1999 Fall Conference of ESK and International Symposium on Ergonomics*, pp. 356-360.

26. The Anthropometric Data of Aging Workers in Taiwan

Mao-Jiun J. Wang, Eric Ming-Yang Wang, and Yu-Cheng Lin

Department of Industrial Engineering and Engineering Management,
National Tsing Hua University
Hsinchu, Taiwan, ROC

ABSTRACT

This paper presents anthropometric data on aging workers in Taiwan. One major anthropometric survey has been recently completed. The new anthropometric database involves static and dynamic dimensions. A direct measurement method was applied to measure 266 static body dimensions and an optical method was employed for measuring 42 dynamic ranges of motions. A total of 11,000 subjects with age ranging from 6 to 65 were measured. A user-friendly computerized query system was designed to help the data retrieval. Here, 58 frequently used aging worker anthropometric data (aged from 45-65) are presented. These data can be applied for the design of workplace equipment and consumer products for aging workers. Eventually, a healthy and comfortable working environment and higher quality working life can be ensured.

Keywords: *Anthropometric data, Aging workers*

INTRODUCTION

Taiwan already is an aged society. The population of aged people of age 65 and above is approaching 9%. As economic growth and technology advancement continue, people tend to have a longer life expectancy and enjoy a better living standard. Thus, it is expected that the aged population will continue to grow. From the human resource management viewpoint, proper utilization and health promotion of the aged and experienced work force is an important issue. According to the International Labor Organization, the criterion for aging worker is age 45 and above. In Taiwan, the proportion of male aging workers is about 26% of the male workforce, and the proportion of female aging workers is about 18% of the female workforce in manufacturing industries. In general, aging workers have decreased physical and mental capabilities as compared to younger workers. One study indicated that the aging worker had a significantly higher mechanical injury accident rate than younger workers in the furniture manufacturing industry [1]. Further, another report indicated that the fatal accident rate for aging workers was also found to be significantly greater than that of younger workers [2]. Thus, it is very important to understand the physical and mental characteristics of aging workers, to properly fit jobs and equipment to the aging workers. Anthropometric data of the aging population is important reference information for achieving ergonomic workplace environment designs.

One large-scale anthropometric survey has been recently completed in Taiwan. The newly developed anthropometric database contains 266 static and 42 dynamic items measured from 11,000 subjects with age ranging from 6 to 65 years old. The database was divided into several subgroups, including workers in the manufacturing industry, army soldiers, college students, senior high school students, junior high school students and elementary school students. The subjects were sampled following the demographic distributions of the corresponding populations [3].

MATERIAL AND METHODS

A 3-D coordinate measuring probe, a digital caliper, and a digital tape measure were used for static measurements. All of these are electronic digital equipment that allow the measured data to be entered into computers automatically with a software interface design. The three instruments and the supporting equipment and software were integrated to smooth the data collection process. With the 3-D coordinate measuring probe, the space coordinates of the surface landmarks were collected, and the relevant dimensions were calculated. The digital caliper was used for measuring the lengths and thickness of body dimensions from the subjects. The digital tape measure was used for measuring body contours or lengths of the curvature. To prevent the input of invalid data due to human or equipment errors, an on-line error-detection program was installed in the system.

A 3-lens infrared motion analyzer was used for measuring the ranges of motions of body segments. To ensure the correctness of the angular values, a number of infrared LEDs (IREDs) were attached on several rigid bodies that were worn by subjects on the relevant limbs when measuring. The positions of the IREDs can be sensed by the CCDs of the motion analyzer and the center line of the measuring body segment can be determined. The angular ranges of motions in degrees were then calculated by comparing the displacement of the center lines before and after a specific motion.

A query program is important for practical and effective use of the database. This program is divided into two main parts, the static and the dynamic anthropometric sections. The static section is used to retrieve the desired data of static dimensions, i.e., length, width, curvature and circumference of a body part. The dynamic section is used to obtain the dynamic data of the ranges of motions of the major body joints.

RESULTS

The anthropometric data of aging workers in Taiwan are presented in Table 26.1. These data that are part of the worker anthropometric data collected from 226 male and 140 female workers. The age and gender distribution follow the worker population distributions in manufacturing industries. As can be seen, due to space limitations, only 58 commonly used static dimensions are provided. The other anthropometric dimensions can be obtained from a CD-ROM retrieval system that is produced by the

Council of Labor Affairs in Taiwan. In addition, the anthropometric data source book of the Chinese people in Taiwan is also published by the Ergonomics Society of Taiwan [4]. Figure 26.1 also illustrates the specifications of the 58 corresponding dimensions in standing and sitting postures.

Table 26.1 The anthropometric data of the aging workers in Taiwan (cont.)

	Dimension (mm)	Male				Female			
		Mean	STD	5th%ile	95th%ile	Mean	STD	5th%ile	95th%ile
1	Weight (kg)	68.2	8.5	54.1	82.2	59.2	7.9	46.2	72.1
2	Stature	1661.3	53.2	1573.8	1748.8	1545.9	52.4	1459.8	1632.0
3	Sellion height	1546.5	53.9	1457.8	1635.2	1429.4	50.9	1345.7	1513.1
4	Menton height	1434.2	52.5	1347.9	1520.5	1326.4	49.1	1245.7	1407.1
5	Left shoulder height	1361.9	47.6	1283.6	1440.2	1264.3	47.0	1186.9	1341.6
6	Left axilla height	1242.0	46.1	1166.1	1317.9	1151.7	45.7	1076.6	1226.8
7	Waist height	968.2	41.8	899.5	1036.9	890.0	43.7	818.2	961.9
8	Iliocristale height	964.8	43.6	893.2	1036.4	914.6	38.6	851.0	978.1
9	Left trochanter height	836.5	43.2	765.4	907.6	790.3	38.6	726.8	853.8
10	Crotch height	691.7	42.7	621.5	761.9	676.6	35.0	619.0	734.2
11	Elbow height, standing	1067.1	38.3	1004.1	1130.1	989.5	39.1	925.2	1053.8
12	Olecranon height	1033.6	38.7	969.9	1097.3	967.2	37.3	905.9	1028.4
13	Wrist height	837.2	33.1	782.8	891.6	784.6	33.4	729.6	839.6
14	Metacarpale III height	744.0	32.5	690.5	797.5	700.8	31.8	648.6	753.1
15	Dactylion height	647.7	30.5	597.6	697.8	610.9	28.9	563.3	658.4
16	Midpoint of kneecap height	440.1	20.8	405.8	474.4	408.3	18.6	377.7	438.9
17	Left lateral malleolus height	66.4	7.7	53.6	79.1	59.1	6.3	48.7	69.6
18	Ankle height	119.8	14.4	96.2	143.4	104.2	13.8	81.6	126.9
19	Sitting height	893.3	28.4	846.7	940.0	839.0	31.8	786.7	891.4
20	Eye height, sitting	779.6	28.7	732.4	826.9	729.1	32.4	675.9	782.3
21	Elbow rest height	261.9	23.4	223.5	300.3	255.6	28.7	208.4	302.9
22	Knee height, sitting	509.9	27.0	465.6	554.3	460.8	20.4	427.2	494.4
23	Popliteal height	395.1	18.0	365.4	424.7	372.3	14.0	349.2	395.4
24	Buttock to knee length	543.4	31.4	491.8	595.1	524.3	25.0	483.2	565.4
25	Elbow to grip length	301.1	27.3	256.2	345.9	265.3	23.1	227.3	303.4
26	Vertical reach	2067.3	77.7	1939.6	2195.1	1900.5	72.7	1781.0	2020.0
27	Vertical grip reach	1943.6	73.3	1823.0	2064.1	1792.1	70.4	1676.4	1907.9
28	Vertical reach, sitting	1298.7	50.5	1215.6	1381.7	1198.7	48.8	1118.5	1279.0
29	Vertical grip reach, sitting	1172.5	44.9	1098.7	1246.4	1087.6	45.7	1012.5	1162.7
30	Arm reach from wall	821.5	37.2	760.4	882.6	759.2	35.1	701.5	817.0

Table 26.1 The anthropometric data of the aging workers in Taiwan (cont.)

	Dimension (mm)	Male				Female			
		Mean	STD	5th%ile	95th%ile	Mean	STD	5th%ile	95th%ile
31	Grip reach from wall	708.3	35.0	650.7	765.9	653.8	32.9	599.7	708.0
32	Head breadth	164.3	8.9	149.7	178.9	164.4	11.7	145.1	183.7
33	Face breadth, zygomatic	133.9	7.2	122.1	145.8	129.1	7.8	116.3	141.9
34	Neck breadth	130.9	12.9	109.8	152.1	117.9	9.5	102.4	133.5
35	Biacromial breadth	373.5	24.2	333.7	413.3	334.9	27.6	289.5	380.4
36	Bideltoid breadth	451.4	24.4	411.3	491.4	426.3	25.8	383.9	468.7
37	Chest breadth at scye	326.7	18.5	296.3	357.2	306.8	17.6	277.9	335.7
38	Biiliocristale breadth	297.7	21.7	262.0	333.3	291.4	25.6	249.3	333.5
39	Bitrochanteric breadth	321.3	17.2	293.0	349.6	331.2	21.3	296.1	366.2
40	Bimalleolar breadth	65.1	4.9	57.1	73.1	61.7	4.8	53.8	69.7
41	Chest depth at scye	212.9	17.0	184.9	240.8	201.0	16.3	174.1	227.8
42	Bust depth	229.9	18.8	199.0	260.9	237.1	22.1	200.8	273.3
43	Waist depth	225.2	27.7	179.6	270.8	222.2	27.7	176.7	267.7
44	Waist depth, sitting	238.6	32.3	185.5	291.7	235.9	31.1	184.7	287.0
45	Buttock depth	225.4	24.7	184.9	266.0	220.9	23.0	183.2	258.7
46	Hand length	182.0	8.9	167.3	196.6	170.4	10.7	152.8	188.1
47	Hand breadth	84.6	5.6	75.5	93.8	77.8	5.1	69.5	86.2
48	Head circumference	579.4	17.2	551.1	607.6	563.6	19.3	531.8	595.4
49	Neck circumference	391.3	25.5	349.3	433.2	355.8	26.7	311.9	399.8
50	Scye circumference	440.6	42.3	371.0	510.2	414.4	46.3	338.3	490.6
51	Shoulder circumference	1108.4	58.7	1011.9	1204.9	1042.4	64.5	936.2	1148.6
52	Chest circumference at scye	941.4	52.3	855.3	1027.4	888.0	59.0	790.9	985.1
53	Waist circumference	864.3	83.5	726.9	1001.7	837.7	94.2	682.8	992.5
54	Waist circumference, sitting	884.9	86.6	742.5	1027.3	848.9	94.5	693.5	1004.2
55	Buttock circumference	940.3	50.7	856.9	1023.7	952.3	56.5	859.3	1045.2
56	Buttock circumference, sitting	1051.0	67.7	939.7	1162.4	1054.0	69.4	939.9	1168.1
57	Knee circumference	365.9	22.5	328.9	402.8	368.5	26.5	324.9	412.2
58	Hand circumference	218.2	17.9	188.9	247.6	204.7	15.7	178.9	230.4

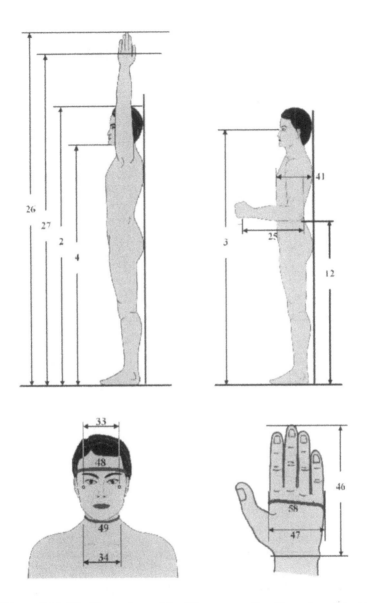

Figure 26.1 The illustrations of the 58 corresponding dimensions (Cont.)

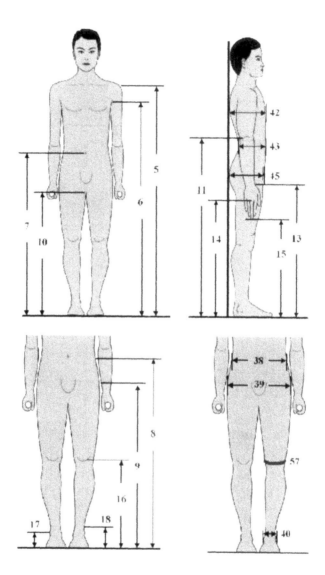

Figure 26.1 The illustrations of the 58 corresponding dimensions (Cont.)

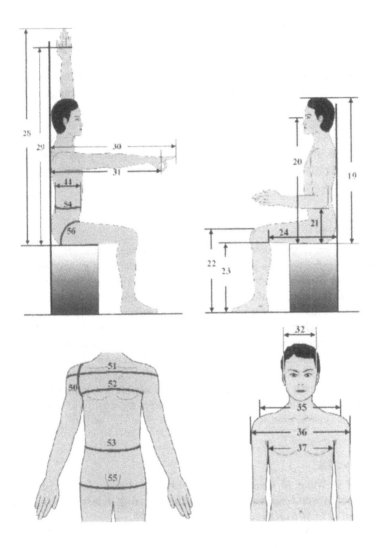

Figure 26.1 The illustrations of the 58 corresponding dimensions (cont.)

DISCUSSION

For the application of the anthropometric database, three major aspects associated with users' body dimensions should be considered: growth trend, age, and gender. It was found that the average statures obtained in this study were 0.24% and 1.39% taller than those of 1986 for the male and female adults, respectively. In other words, Taiwanese workers have grown markedly in the past decade. This trend should be also considered for future applications and studies.

For the growing trend of globalization, it is necessary that the anthropometric data of different ethnic populations should be referred to when producing products for different target markets of customers. In addition, many developed countries also have the phenomenon of the aging society. Thus, the availability of the anthropometric data of the aging people and anthropometric data of different ethnic groups is also essential.

The query program of the anthropometric database is an open-structured software system, with which the new anthropometric data can be added and updated whenever necessary. This program is simple, easy to use, and provides practical information. These characteristics ensure the usability of the program and assist the users to access the desired data.

CONCLUSION

As the population of aging workers continuously increases in the workplace, it is very important to understand the physical and mental characteristics and limitations of aging workers in order to better fit them to jobs and equipment. The anthropometric data are very fundamental information for ergonomic jobs and workplace designs for aging workers. From the recently completed anthropometric database in Taiwan, there are 266 static and 42 dynamic anthropometric dimensions of aging worker populations are available. They can be used for various facility and product designs. In this paper, 58 commonly used static dimensions are provided. Other information can be obtained from the published anthropometric data book or the CD-ROM version of the data retrieval system for the worker population in Taiwan. It is our ultimate desire that through the use of this anthropometric information, the quality of working life for aging workers can be enhanced.

REFERENCES

1. Ma, W.S., Wang, M.J. and Chou, F.S., 1991, Evaluating the mechanical injury problem in wood-bamboo furniture manufacturing industry. In *International Journal of Industrial Ergonomics*, 7, pp. 347-355.
2. Chi, C.F. and Wu, M.L., 1997, Effects of age and occupation on occupational fatality rates. In *Safety Science*, 27, pp. 1-17.
3. Wang, E.M.Y., Wang, M.J., Yeh, W.Y., Shih, Y.C. and Lin, Y.C., 1999, Development of anthropometric work environment for Taiwanese workers. In *International Journal of Industrial Ergonomics*, 23, pp. 3-8.
4. Wang, M.J., Wang, E.M.Y. and Lin, Y.C., 2002, Anthropometric Data Book of the Chinese People in Taiwan, The Ergonomics Society of Taiwan.

27. Development of a Work Support Tool for the Old with Work Postures as an Index

Masahiro Shibuya[1], Koki Mikami[1], Mitsuyuki Kawakami[2] and Masaharu Kumashiro[3]

[1] Hokkaido Institute of Technology, Japan,
[2] Tokyo Metropolitan Institute of Technology, Japan
[3] University of Occupational and Environmental Health, Japan

ABSTRACT

In Japan "Fewer children and aging" has become a problem. As workers advance in age, they are forced to conduct physically excessive work because of weaking in their strength. In order to reduce the load of manual labor, it is necessary to improve conditions by evaluating quantitatively the load of work postures and finding out improvement points within the factory. The author et al have been developing a work posture burden evaluation system to reduce the physical (muscular) load of workers. We investigated work postures appearing frequently during work, EMG-measured each work posture, and gathered the results in a database. Then we provided the database with a visual interface, and created a system which can display both work postures in the 3D model and their burden evaluation indices. In this paper are described how to actualize the system and the outlines of our present trial one.

Keywords: *Aging, Elderly worker, Work burden*

INTRODUCTION

In our country the population rate of people over 65 was more than 17% in 2000, and it will be more than 22% in 2010 and more than 26% in 2020. Japan is on the way to a "super aging" society. This rapid aging produces various influences upon industries. For example, many older people have already got durable consumer goods such as household electrical appliances and cars, which fact necessitates diversified small-quantity production of products to meet the needs of consumers. Further, the needs of older people are thought to shift from goods to service.

In production fields, it is thought that productivity will be caused to lower greatly by lowering of work efficiency due to aging, lowering of productive capacity due to decrease in annual working days, and so on. Moreover, in large-scale manufacturers, which have shifted their production overseas and reorganized their domestic production, a new problem, a changeover from conventional simple assembly factory workers to workers with many skills, has arisen. And also, fewer young people have been finding employment in the primary and secondary industries, and securing successors after simultaneous retirement of skilled older workers is becoming a problem.

Enterprises cannot expect to employ younger people as before, and they are compelled to contrive to employ women or older people instead in order to maintain the present productivity. However, in small and medium enterprises man-machine labor remains unchanged, and many kinds of tasks requiring excessive physical load exits. For instance, the load is caused by uneasy work postures, unsafe work, the deterioration of workers' physical powers due to their aging. It has become increasingly important for enterprises with more older workers to reduce the load of their physical work.

Considering these present social conditions, we will need a system which helps us to perform work improvement to reduce the physical load of older workers and create a workplace friendly to workers. The purpose of this study is to develop a support tool for work improvement which anybody can use easily when people in the workplace, in a body, try to improve their working environment and work methods and create a comfortable workplace.

In this paper are described our view of encoding of work postures, how to actualize the system, how to calculate the evaluation indices of work postures load, the outlines and utilization of 3D CAD, and our present trial system itself.

SYSTEM CONCEPT

In order to have people of middle and advanced age show their full ability and improve productivity further, devising logical work methods is necessary. We need to mechanize much of the process requiring rough work using support tools. We can reduce their workload by devising heavy-object handling and by improving work postures. And also standardizing "improvement knowledge" and sharing it enables us to make wide use of information and knowledge.

Thereupon the author et al turned our attention to work postures which are easy to observe without disturbing workers. We thought that using work postures as an index would be an effective way from the viewpoint of analysis of prevention of disorders such as lumbagos, safety, working patterns and so on. We made it possible to calculate a burden index from the muscular load corresponding to each work posture so that we could indicate improvement points and measure the effect of improvement. Further, we thought it necessary to process the data of the accumulated know-how, specifically intuition and experience, make a database easy to use, utilize it for work improvement. Fig. 27.1 shows the image of the usage of this system.

Our Systems

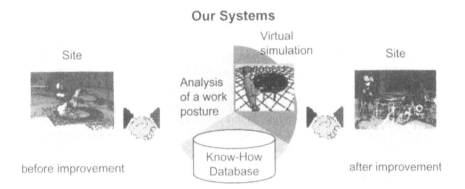

Figure 27.1 Concept of System

SYSTEM DESIGN

Work postures and burden evaluation indices

We conducted an investigation at 31 worksites in 6 factories, and decided on a code for each work posture. Each work posture is represented by a 6-figure code which symbolizes the upper limbs, the loins, anteflexion of the uterus, the lower limbs, the feet and "holding a heavy object". The code is a combination of figures and a letter. In EMG (surface-electromyography) tests, we gauged muscle strength for 30 seconds at 10 points, namely M. deltoideus, M. erector spinae L5, M.vastus medialis, M. gastrocnemius and M. tibialis anterior in the right and the left sides, in 1096 kinds of work postures.

It was decided that the total of the maximum muscle strength rations (30 sec, 100Hz, the ratio the average amplitude of amplitude integral values: the average amplitude of maximum muscle strength values) of the 10 points be the quantitative evaluation index of each of the work postures. The burden evaluation index of a task is the total of the burden evaluation indices of work postures during the task multiplied by their occurrence rates. The burden evaluation index of a process is the total of the burden evaluation indices of tasks multiplied by their occurrence rates.

Computer system design

The system we are developing now is designed so that we can find clues to work improvement by inputting work postures obtained using Snap-Reading Method (a method of observing instantaneously the conditions of an object at random time) and several parameters, and evaluating them. Further, it is important to lessen the input work as much as possible because the users are on-site workers. We are developing the system in consideration of the above.

The system design we are now developing is shown in Fig. 27.2. The rounded squares show the actions performed at the worksite, the squares the tools we are developing, and the ellipses the data defined by this system. The flow of data and control is shown by the arrows.

There are two ways to input work postures. One is the use of a work-posture input tool shown on the left of Fig. 27.3. The other is automatic input, by a motion capture device, of the markers as large as a ping-pong ball affixed on a worker (on the right). The tool input is performed by choosing an appropriate item from each of the six list boxes for the upper limbs, the loins, the upper part of the body, the lower limbs, the feet and weight load on the window. In this input we have to input many times work postures obtained through work sampling when the sampling interval is set short.

Therefore we made a study of inputting work postures automatically with a motion capture device to automate the input work. We used Motion Graber II by OYOUKEISOKU. This device makes it possible to output the positions of the markers affixed on the body in 3D coordinates. The output 3D data cannot be used in their original condition, so they are converted to 6-digit work posture codes using the data analysis tool. The output data have missing links, that is to say unmeasurable portions, because the worker walks out of the camera view or goes behind some object. Consequently it is necessary to conjecture the missing part from the context.

The input data of 6-digit work posture codes are utilized for calculating a quantitative burden evaluation index from the muscular load corresponding to each work posture. Burden evaluation indices are obtained from the input work postures using our burden evaluation index tool of work postures. This tool makes it possible to understand improvement effect quantitatively by comparing the indices before and after improvement.

Simulation using 3D CAD Software

Virtual simulation with a human model is made possible by processing the data of work posture codes transmitted to the 3D CAD software. Introduction of support equipment can reduce the physical load from carriage work and unnatural work postures with heavy load on a part of the body such as the loins or the head. Therefore CAD software for processing the support equipment and the human model simultaneously is needed.

We supplemented IGRIP by DELMIA with ERGO Module to program 3D models of the human body. IGRIP is a software system which is used in off-line programming of robots, space station assembling etc.

The addition of ERGO Module, which provided the system with the human model of 86 degrees of freedom and kinematics, made it possible to simulate human motions.

In order to make the human body of DELMIA/ERGO take each individual posture determined in advance, we developed connecting driver and interface software using C++Builder. The postures defined by the program are stored in DELMIA/ERGO in the unit "posture". Human motions were made possible by defining the sequence which defined plural postures and their motion time. Fig. 27.4 shows a simulation of welding work using support equipment defined by the posture data transmitted to the CAD system. The previous work posture is on the left and the one after introduction of support equipment is on the right.

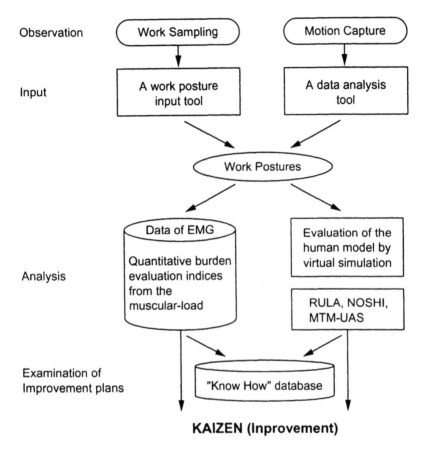

Figure 27.2 System designe

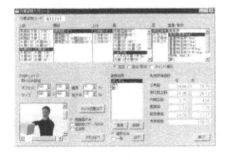

(a) work posture input tool

(b) 12 markers are put on a worker

Figure 27.3 How to input a work posture

(a) before improvement   (b) after improvement

Figure 27.4 Evaluation of the human model by virtual simulation

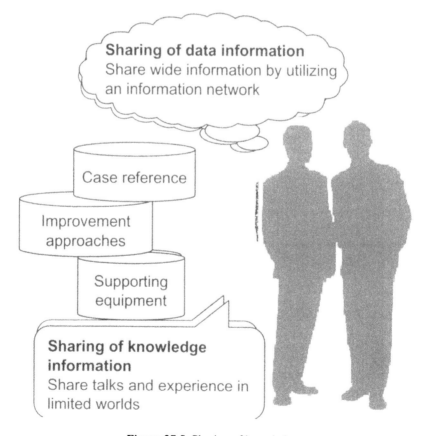

Figure 27.5 Sharing of knowledge

Sharing of work improvement information with an information network

When we set to work improvement actually, there are cases where past improvement cases and know-how are helpful. We need to share workers' talks and experience in limited worlds such as teams and wide information by utilizing an information network (See Fig. 27.5). Accumulating personal experiences and experience as know-how in a database is necessary for that sharing. Further, we need to store there improvement methods such as "JIT (Just In Time)" and support equipment such as a balancer in an easily accessible kind of form.

Connection of this know-how database with the Internet enables anyone to get information he wants, anytime, anywhere.

CONCLUSION

In this study we developed a quantitative work posture burden evaluation system on the basis of Electromyography for the purpose of reducing workers' physical (muscular) load at the production site, and evaluated the system. In requires only input of work postures, and has many uses in industries other than the manufacturing industry. The system is thought to be useful in the field of walfare for the elderly and so on.

The fusion of DELMIA/ERGO and the system we are now developing made it possible to convert input work postures to 3D CAD models and simulate their motions. The conversion to 3D CAD models made possible advance simulation of work with newly developed support equipment. Moreover, the fusion enabled us to use the functions of DELMIA/ERGO such as the calculation of energy consumption calories and the analysis of MTM-UAS cycle time.

By making this system correspond to information devices, such as wire-less LAN and "Palm" computers, which have made rapid progress in recent years, it will be an improvement support system accessible to everyone, anytime, anywhere.

28. A Study on the Usability of Mobile Phones for the Elderly

Kwan Suk Lee, Bohyun Kim

*Department of Information and Industrial Engineering, Hongik Univ,
Seoul, Korea*

ABSTRACT

The elderly will gradually form a very significant proportion of the telecommunications market as the population ages more quickly. However, older people are experiencing difficulty in using mobile phones because most mobile phones are designed for the young generation. Access to telecommunications offers the elderly more independence, mobility and better quality of life. The aim of this study was to determine the needs of the elderly in using mobile phones and find ways to improve the usability of mobile phones for the elderly. It was found that people over 50 years old normally have trouble using the functions of mobile phones used in Korea; most of them could use only the basic functions.

Keywords: *Usability, Mobile phone, Elderly*

INTRODUCTION

At present, the mobile phone most commonly used in Korea provides a variety of functions such as news, checking e-mail service, games, stock information, etc. Yet most of these functions are designed for the younger generation, who can easily handle complicated electronic devices. Many functions added to mobile phone have been causing difficulty in using the phones, not only for elderly people, but for middle-aged people, as well.

In fact, most of the elderly tend to use only the basic functions, such as making/answering calls. They rarely use the many other useful functions of a mobile phone. This implies that usability of mobile phones does not consider all users' needs in the design process, especially the needs of the elderly.

In 2000, Korea became an aging society, in which people over 65 years old are more than 7% of the population. This number will increase to 10% by the year 2010 and 19% by the year 2030. The incremental rate of aging shows a rapid increase compared to other advanced countries. It took 115 years for France, 85 years for Sweden, 47 years for the United Kingdom and 24 years for Japan to become aged societies (14% of the population over 65 years old) from aging society (7% of the population over 65 years old). It is expected that Korea might take 22 years to become an aged society, even faster than Japan [1].

Elderly people form a significant proportion of the telecommunications market, yet the industry has largely ignored their needs [3]. The individual variation in the functional ability of people of different ages is, however, very large. Increasing disabilities and functional limitations are not just issues for those in their 60s or 70s. According to available research, people who are reaching their fifties are having more problems than they had in the past with some functional abilities [2]. Therefore, we decided to include people who are in their fifties as well as people over 60 years old in our study, since they will be potential consumers in the future.

This study surveyed the current problems of people in their fifties and over in using a mobile phone and analyzed the problems by usability tests.

MATERIAL AND METHODS

Survey: Our survey was conducted to find the subjective satisfaction with design factors of the mobile phone. 120 participants were divided into 4 age groups from the 20s to the 50s. The survey was conducted using a person-to-person questionnaire survey. The questionnaire included general information on the user; frequency of use of each function, divided into five categories; subjective satisfaction with six design factors and five menus. All participants had at least two years' experience in using a mobile phone, and were using a mobile phone that was produced in Korea in the last two years. This can reduce the effect of using different functions of different mobile phones.

Usability test: 10 younger adults aged from 21 to 29 years ($\underline{M}$=25.6, $\underline{SD}$=2.46) and 10 older adults aged from 51 to 58 years ($\underline{M}$=53.8, $\underline{SD}$=2.1) participated in the test. The test was composed of three kinds of tasks, i.e., adjust volume of the phone, input a given phone number and name into the phone, and find the input number. Performance time and error were measured for each task to compare the performance between younger adults and older adults. The mobile phone used in the test was the latest model. All subjects had more than two years of experience in using a mobile phone produced during the past year in Korea. However, they had no experience with the mobile phone used in the experiment and were given instruction and a manual by a trained instructor.

RESULTS

Survey

Frequency of Use of the Mobile Phone per Day by Each Age Group

Figure 28.1 shows the number of functions used in a day for different age groups. There is little difference in making/answering a call for all ages, and the frequency of use of other functions decreases as age increases. Particularly for those in their 50s, the frequency of using functions such as sending text massages and mobile Internet service shows that they seldom use these functions. The reasons why they do not use these useful functions were that they thought 'it is annoying and unnecessary' (63.3% of non-users of functions) and 'do not know how to use' (36.6% of non-users).

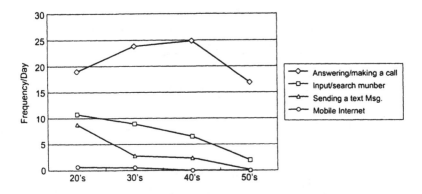

Figure 28.1 Frequency of use of functions

Subjective Satisfaction with Six Design Factors of the Mobile Phone

Figure 28.2 shows the subjective satisfaction with six design factors related to the size of the mobile phone. The factors were chosen to identify the characteristics of those age 50 and older based on the fact that there are physiological differences between age groups, such as vision, hearing, tactile sensation, etc.[2]. The questionnaire was composed of categories such as size of the mobile phone, buttons, LCD, letters on both LCD and buttons, and the volume of the speaker. All participants were asked to give marks for each question. The point range is from 1 to 7. Higher numbers mean that they wanted a bigger size or higher volume. The lines on the graph show the average points for the six design factors in each age group.

There is a slight difference in the favored sizes for the mobile phone and the LCD in all age groups. The average points of the subjective satisfaction with the LCD do not show a large difference between the 20s and the 50s. On the other hand, the results for the size of letters on the keypad and LCD show that those in their 50s are having difficulty in reading small letters. It implies that those in their 20s want a large LCD so as to see more information at a time. On the whole, the 20s and 30s have similar results, and it is the same as those for the 40s and 50s.

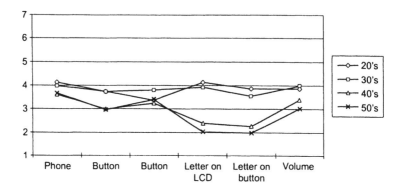

Figure 28.2 Subjective satisfaction with six design factors

Subjective Satisfaction with the Menus of the Mobile Phone

Figure 28.3 consists of five mobile phone menus, which are making/answering a call, input/search numbers, sending text messages, using mobile Internet service, and using mobile phone options. Subjects were asked to rate their subjective satisfaction with these five menus in their daily use of the mobile phone. The point range is from 1 to 7. Lower numbers mean that they were not satisfied with the menu and they wanted it to be redesigned for easy use. The lines on the graph show the average points for the five menus in each age group.

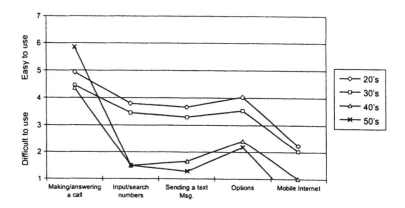

Figure 28.3 Subjective satisfaction with the menu

The results indicate that older people have more trouble in using menu functions in general. Apart from making/answering a call, all the other results show that using mobile phone menus becomes more uncomfortable as a person become older. On the contrary, the subjective satisfaction points for making/answering a call indicate that those in their 50s gave the highest points out of all the age groups. It can be inferred that the 50s have more satisfaction with the making/answering a call function than the other menus of the mobile phone.

Subjective satisfaction with the mobile Internet service appears to be the lowest for all age groups. 54.5% of users of the mobile Internet service said that it was very confusing when they navigated. This was due to its complicated menu operations. 36.7% of the users said that it took too much time to connect to the service. For input/searching numbers, 33.3% of those in their 50s were using a quick dial function. This function is supposed to be used after they input the phone numbers into the mobile phone. But only 43.3% of these users could do it by themselves and the rest needed help when they wanted to input a phone number.

Comparison of Requirements between Age Groups

All subjects were asked which mobile function was difficult for them to use. Table 28.1 shows the rankings and the percentages of each age group's requirements for the mobile phone. 'Ease of searching menus' and 'Good array of Korean alphabet on keypad' score high in all age groups. It suggests that usability of the mobile phone is not just an issue for older people, but also is a problem for young people. Most mobile phones used in Korea at present have many functions. The usability needs to be studied in the design process. Also, it is shown as age increases, the number of requirements and percentages of users asking for requirements decrease. This is related to the result that was shown in Figure 28.1, where the frequency of use of functions decreased with increasing age.

Table 28.1 Comparison of requirements between age groups

Age groups / Requirements	20's Rank	20's %	30's Rank	30's %	40's Rank	40's %	50's Rank	50's %
Good array of Korean alphabet on keypad	5	30	1	26.7	2	23.3	7	3.3
Ease of adjusting volume	7	23.3	9	6.7	6	13.3	2	16.7
Ease of searching menus	2	43.3	2	23.3	2	26.6	1	20
Light weight and thin shape in design	3	36.7	6	16.7	1	30	6	6.7
Long battery durability	8	20	14	3.3	8	10	3	13.3
Size of letters on LCD	-	-	9	6.7	6	13.3	3	13.3
Size and space of buttons	-	-	3	20	4	20	3	13.3
Design diversity	6	26.7	3	20	5	16.7		-
Reduction of navigation time in using mobile Internet service	11	10	9	6.7	-	-	-	-
Size of LCD	3	36.7	9	6.7	-	-	-	-
Additional various function	1	53	3	20	-	-	-	-

Usability test

Test Objectives

An objective for this test was to identify problems of older adults in using a mobile phone by comparing performance time and errors between those in their 20s and the 50s on three tasks. The aim was to give design proposals that enhance efficiency, productivity and user satisfaction of the elderly in using a mobile phone.

Design for Usability Test

Subjects were asked to complete three different tasks. The first task was adjusting the volume of the ringing sound, and the second task was inputting a given number and a name into the experiment mobile phone. The final task was searching for a phone number that had been stored in the second task. To perform the test, subjects had to complete 5 steps for task 1, 8 steps for task 2, and 7 steps for task 3. The main instructions for the tasks are listed in Table 28.2.

Table 28.2 The main instruction of the three tasks

	Task 1	Task 2	Task 3
Step 1	Press 'Menu' button on the phone	Press a given number and 'OK' button on the phone	Press 'Phone book' button on the phone
Step 2	Select 'Ring/Vibrate' by pressing the button on the phone	Select 'New Name' by pressing the button on the phone	Select 'Quick Search By Group' by pressing the button on the phone
Step 3	Select 'Ringer Volume' by pressing the button on the phone	Select Korean alphabet keys for input a given name by pressing the buttons on the phone	Select 'Group Name' with buttons on the phone
Step 4	Adjust volume with buttons on the phone	Press 'OK' button on the phone	Press 'OK' button on the phone
Step 5	Press 'OK' button on the phone	Select 'Group Name' with buttons on the phone	Select the name with buttons on the phone
Step 6	-	Press 'OK' button on the phone	Press 'OK' button on the phone
Step 7	-	Select 'Cell Phone' by pressing the button on the phone	Press 'Send' button
Step 8	-	Press 'OK' button on the phone	-

Test Results

Figure 28.4 indicates that older subjects are much slower than younger subjects in performance time. Older subjects had a poor understanding of the manual, and thus more help was needed when they performed the operation. They also had trouble reading the letters on the LCD and buttons, which made it very difficult to perform the tasks. It was also found that older subjects tended to read through all the options on the LCD before making a selection, while younger subjects tended to choose an option immediately.

Older subjects also made many more mistakes in pressing a button or making a choice because of bad vision or lower tactile senses. Older adults were more dependent on reading when they made a choice, while younger subjects tended to do it even without reading the manual. A part of this difference could be explained by the fact that most of the younger subjects had experiences in electronic devices while the older subjects did not have such experience.

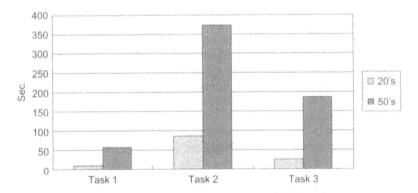

Figure 28.4 Performance mean time

Table 28.3 shows the performance time and error times during the test. Results indicate that those in their 50s take about 4 to 7 times longer than those in their 20s to complete the tasks. Considering that task 2 and task 3 have similar operations, it could be speculated that there was a learning curve, since mean performance time of the 20s and the 50s was reduced by 67% and 50% respectively in task 3 compared to the mean time of task 2.

The results of Standard Deviation (SD) for those in their 50s, which is much bigger than for those in their 20's, are supported by the fact that individual differences in various cognitive functions become greater with age [4].

Table 28.3 Performance time and times of error

		Mean time	SD	P-value (0.05)	Mean errors
Task 1	20's	10.6s	3.11s	2.98E-13	0.4
	50's	58.8s	7.52s		2.1
Task 2	20's	86.8s	30.11s	7.2E-06	1.3
	50's	374.5s	143.13s		7.4
Task 3	20's	26.6s	9.78s	0.000647	0.3
	50's	188.2s	123.74s		5

CONCLUSION

More advanced equipment often requires user knowledge and skills, and thus people who were already experiencing difficulty can actually be further isolated by technical advances. There are also intrinsic problems in the nature of new technology.

As products change rapidly, shrinking in size and cramming more and more features into the software, engineers focus on getting the chip inside or modifying the software while the product is being developed rather than considering users. Devices for people with special needs tend to be obsolete, non-competitive and difficult to maintain [3]. However, it is important to consider the usability of the mobile phone based on all users' needs rather than just to develop a new technology because it is becoming a necessity of daily life.

Convenience of access to telecommunications provides a better life for the elderly. Development of functions for the elderly, e.g., alarm systems, appears to be very effective and helpful in coping with their safety problems (Bouma, H, Graafmans, J A.M. 1992, 47-48). Also, aspects of mobile phone function will enhance the activity range of older people and affect their independence and financial capability.

REFERENCES

1. Korea National Statistical office, 2000, Korea in figures, (http://www.nso.go.kr/report/data/svgg2000.htm).
2. Ekberg, J., 1999, General issues regarding ageing and technology. In *INCLUDE project*.
3. Gill, J., 1994, The Forgotten Millions-Access o telecommunications for people with disabilities, International Telecommunication Union.
4. Ekberg, J., 1999, Factors affecting uptake and use of technology. In *INCLUDE project*.
5. Bias, R.G. and Mayhew, D.J., 1994, Cost-Justifying Usability, (London: Academic Press).
6. Roe, P., Gjoderum, J., Hypponen, H., Nordby, K., Guud, S.E., Ekberg, J. and Martin, M., 2000, Guidelines-Booklet on Mobile Phones. In *COST 219bis Guidebook*.

7. Yang Lam, C., Tsang, R., Soparkar, S. and Santis, S.D., The Design of an ergonomic mobile phone,
(http:// peach.mie. toronto.ca/courses/ mie240/html/group44.html).
8. Lindstrom, J.I. and Martin, M., 1996, TELEMATICS project 1109, Emergency Services and Alarm Systems. In *Telecommunication for all*, p. 173.
9. Hypponen, H., 1997, Disability and ageing, Stakes.
10. Schwartz, J., 1999, Making Cell phones Disabled-Friendly. In *Washington Post, July 14*, P.E1.
11. COST 219bis, Human Aspects of Telecommunications for Disabled and Older People, Swedish Handicap Institute, p.75, 1999.
12. Jordan, P.W. and Thomas, B., 1996, Usability Evaluation in Industry, (Taylor & Francis).

29. Ergonomics Problems in Job Redesign for Small-to-Medium Sized Factories

Tetsuya Hasegawa[1], Masaharu Kumashiro[2]

[1] *Kyushu School of Engineering, Kinki University, Fukuoka, Japan*
[2] *Department of Ergonomics, UOEH, Kitakyushu, Japan*

ABSTRACT

At many workshops, working systems are not necessarily designed for the benefit of older workers. In this paper, factors that would determine the ease of application of the technique to problems of how to create a good working environment for middle and advanced aged workers were studied based on the two cases.

As a results, it is concluded that the basic conditions for work improvement include, in terms of the company, is to establish an in-house system that makes effective use of knowledge or techniques for production management. In terms of industry, it is firstly to establish a supporting system in which engineers can learn production management. Secondly, to provide an opportunities to visit other companies and factories and to exchange information, and finally to make available information regarding the examples and suggestions for work improvement.

Keywords: *Job redesign, Small-to-medium sized factory, ERGOMA approach*

INTRODUCTION

Japan has a rapidly aging population and therefore urgently needs a working environment in which the aged people feel comfortable to work. With advances in age, the legs and body are enfeebled while visual and other sensory functions are lessened. Many working people want to continue working even when they are far advanced in age. Reduction of physical work load is important for the employment of older workers. At many workshops, working systems are not necessarily designed for the benefit of older workers. The will of the management to improve working systems and the united efforts of the organization including the workers themselves are necessary to cope with this problem.

The Association of Employment Development for Senior Citizens promotes research on issues of employment for aged workers. As part of its activities, collaborative research is being carried out in cooperation with private corporations on development of employment opportunities for older workers. The Association provides half of the research expenditure for these joint projects.

In these research projects, the ERGOMA [1] approach has been traditionally employed. This is a technique that analyzes a situation from three viewpoints: human engineering, industrial engineering, and occupational health. Successful results have been achieved so far from the application of this technique to problems of how to create a good working environment for middle and advanced aged workers.

It has been observed from past applications of this technique that there are two types of working places where the approach is effective: one is where the technique works well, and the other is where it takes time to start working. Therefore, factors that would determine the ease of application of this technique were studied based on the two cases.

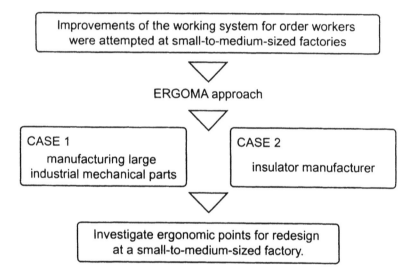

Figure 29.1 Flow of the research

CASE 1

Outline of work

The first case of successful application of ERGOMA approach is a company manufacturing large-size industrial mechanical parts. It specializes in integrated production process, from processing to assembling products such as steel-manufacturing machines, tire production equipment, tire molds, and other general-purpose machine tools. The company used 35 workers in line (mean age 44.0 years).

Figure 29.2 Before job redesign (Assembly process with crane operation)

Method of approach

The company had three major concerns: Firstly, how to make the middle-and advanced-aged workers cope with numerically controlled machine tools, which are being introduced in greater number. Secondly, how to hand down the expertise, knowledge, and skills of experienced workers to younger generations, and finally, how to create a working environment convenient for the elderly.

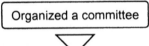

Organized a committee

Worklaod
   Activity sampling,
   Working pastures (OWAS)
   OFF, Heart rate
   Feelings of fatigue

Working environment
   Thermal environment, Noise level, Luminance level

Occurrence of unsafe practiced in the past

Operativity of ceiling cranes

Figure 29.3 The job redesign approach for the factory manufacturing industrial mechanical Parts

For the issue of working environment sympathetic to the elderly workers, the current condition was investigated from four standpoints: (1) workload; (2) working environment; (3) occurrence of past unsafe practices as observed from work log books; and (4) operativity of ceiling cranes, thereby attempting to find solutions to improve the working environment.

As part of the job redesign, activity sampling, working postures, workload and the working environment were investigated.

The workers' activity and working postures were observed and sampled every thirty seconds. The Ovako Working Posture Analyzing System (OWAS) (2) method was used to identify and evaluate working postures. The critical flicker fusion frequency (CFF), heart rate and feelings of fatigue (30 items concerning the subjective feelings of fatigue, as published by The Japan Association of Industrial Health) were examined to estimate the workload. The thermal environment, the noise level, and the luminance level also were measured.

It turned out that ceiling cranes were the source of much unsafe conduct. As they were very useful and helpful in making effective use of the workspace, they had been very frequently used.

Results

Trying to eliminate unsafe conduct with respect to the ceiling cranes, the study suggested change of the ceiling crane system from conventional cranes to two ceiling cranes, large and small, capable of adjusting the vertical moving speed of the hook. The new pair of cranes proved to be a success, particularly, in providing good balance during the lifting of products. Some lighting was also installed onto the cranes as illumination on the workspace decreased as the cranes moved.

The drawing table was also improved so that it could be accommodate various sizes of drawings, again more lighting equipment was installed.

Figure 29.4 Crane improvement

CASE 2

Outline of work

The company used 105 workers (mean age 47.3 years). By age distribution, 34.4% of the employees were 40 to 49 years old, 31.3% 50 to 59, and 7.3% above 60. The factory undertakes the grinding and mixing of raw materials, forming, glazing, heat treating, packing and shipping.

The factory makes about 400 different types of insulators with a weight of between 10 g and 80 kg. Materials transportation was done by carrying or by hand cart. There was no air conditioning, only electric fans in the factory, since dry air is unsuitable for the materials.

The working hours were from 8:00 to 16:50 with lunch time from 12:00 to 12:50.

Young people in Japan seeking employment in industrial fields feel a strong aversion to dirty, hard and dangerous jobs. Therefore this company always encounters difficulties in employing young workers. The supervising staff had little chance to learn how the manufacture occurred in this factory, because they had to spend most of their time in production work.

The company had revised their wage and personnel-management systems, extended the age limit and proposed a continuing employment program for the present workers who would continue beyond retirement. However, none of the older workers at this company would accept to work beyond retirement. It seems that they flinched from the proposition, because the job involved hard physical labor and they were reluctant to ruin their health by continued engagement in damaging work.

Figure 29.5 Insulator samples

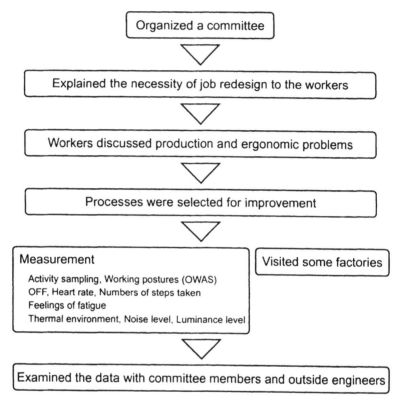

Figure 29.6 The job redesign approach for the factory manufacturing insulators

Method of approach

The committee for the promotion of the employment of middle-aged and older workers was composed of managerial staff and university researchers. To explain the necessity of job redesign to the workers, they discussed the production and ergonomic problems of each process, and representatives of each process presented the results to the committee. These results contributed to the committee members better understanding of the production methods. To seek data necessary for improvement and to determine the method of investigation, preliminary research was carried out by the committee. Thus, the processes of grinding, blending and milling, ageing and tempering, were picked out for improvement.

Activity sampling, working postures, the workload, and the working environment were investigated. The workers' activity and working postures were observed using the OWAS method. The critical flicker fusion frequency (CFF), heart rate, number of steps taken, and feelings of fatigue were examined. The thermal environment, the noise level, and the luminance level also were measured.

The results achieved by this project were evaluated using the same measurement.

Results

Analysis of currently existing problems

Unnatural working postures such as bending forward or twisting from the waist were frequently observed in workers.

Work activity level was low. In the grinding process, the worker often took a rest and a drink of water because of the heavy physical work with a shovel in hot temperatures. In the ageing and tempering process, the setting and re-setting of the machine for different jobs were frequent because of the many types of production involved.

In the blending and milling processes, auxiliary operation in running the mills, such as moving heavy trolleys, opening the cap and pouring water into the mills, were also frequent. The worker had to go up and down narrow stairs many times to check the operating condition of the mills

Every worker felt pain in his back, and some workers felt tired in their legs and stiff in their shoulders after work. These seemed to reflect these unnatural working postures.

Plans for improvement

The problems which existed may be summarized as;

1. Work entailed unnatural working postures and heavy physical work
2. Problems arose from the physical environment

Workers and staff had little awareness of these problems. Many workers had worked in other ceramics factories in this area so they took such working conditions for granted. There were few chances to visit other industries. We explained our analysis of the current working environment to give the workers a better understanding of ergonomics. Through this process, awareness of the seriousness of the problems gradually grew.

The suggested course of action for improvement was;

1. Improving working postures and lowering the workload
2. Improving the working environment

Introduction of improvement plans

With the help of ceramics engineers from outside the company, suggestion for improvement of the work system, factory facilities and equipment were tabled for discussion. At this stage, it was difficult for the engineers to gain an understanding of ergonomics. They had never applied ergonomic principles to making facilities and equipment. Through the meetings, engineers gradually came to understand the principles of ergonomics.

The proposals were as follows;

1. Stop manual shovelling: a large-sized hopper was set within the grinding process so the worker can transfer the materials in a power shovel.
2. A ceiling crane and large sized trolleys were introduced into the mixing process. With these, the number of times that the worker pushes the trolley is decreased, and the
3. frequency of trips up-and-down stairs is decreased.
4. Improvement of water tanks: using water tanks whose capacity was decided according to the size of each mill, the frequency of water bulb control and bulb control error was reduced.
5. Introduction of a lifter: some commercial lift equipment was introduced in the ageing and tempering processes. the previous bending posture was eliminated.
6. Improvement of work environment: ceiling height was increased.
7. Improvement of work organization: the supervisory staff can now find time to do their own jobs.

Joint research brought unexpected benefits, eventually with various ideas put into practice, such as the improvement of the drawing table, which was originally not planned. These improvement stimulated workers greatly. Employees' motivation for the work appears to have been very much improved.

Figure 29.7 Before and after trolley improvement

Figure 29.8 Before and after lifter introduction

DISCUSSION

When these collaborative researches are conducted, specific work would normally be necessary. This may include identification of problems currently occurring in the present processes and determination of which problems to be addressed as targets of improvement. In addition, as existing data is not sufficient, past data needs to be newly investigated, compiled, and analyzed in order to pick up issues worth being called "problems." This type of work requires personnel with knowledge and experience on production management to some extent. In case 1, fortunately the company had the very person, which is why they were able to pursue the joint research very swiftly and smoothly. The background of this person was important. He originally worked for a large firm and gained education and training on production management and eventually acquired practical knowledge and experience. Later, after retirement, he found his second career at this manufacturer. His knowledge and experience made a great contribution to the joint research project.

Essential requirements for work improvement and its realization include knowledge of ergonomics, and the funding for actual implementation of improvement measures. Equally important is whether or not the knowledge of industrial engineering, such as production management, which serves as a basis for manufacture, is effectively used within the company. It is, however, difficult for small and medium-sized businesses to raise qualified engineers by themselves and employ them.

Supervisors in small companies have little time to devote to managerial work because they often have to act as production workers. It is of primary importance that the staff secure the time to devote themselves to managerial work [3].

CONCLUSION

In small companies, the supervisors have little time to devote themselves to managerial work, since they often have to act as production workers. It is of primary importance for the staff to secure time to devote themselves to managerial work.

Based on the above discussion, it is concluded that the basic conditions for work improvement include, in terms of the company, is to establish an in-house system that makes effective use of knowledge or techniques for production management. In terms of industry, it is firstly to establish a supporting system in which engineers can learn production management. Secondly, to provide an opportunities to visit other companies and factories and to exchange information, and finally to make available information regarding the examples and suggestions for work improvement.

REFERENCES

1. Kumashiro, M., 1987, Work load, postures and job redesign: An ergonomic and industrial management (ERGOMA) approach. *New methods in applied ergonomics*, edited by Wilson, J.R., Corlett, E.N. and Manenica, L., pp.247-252.
2. Louhevaara, V., Suurnakki, T., 1992, OWAS: a method for the evaluation of postural load during work, Institute of Occupational Health Center for Occupational Safety Helsinki,.
3. Hasegawa, T. and Matsumoto, K., 1995, Job redesign for elderly workers in a small-medium-sized noodle manufacturing. *The paths to Productive Aging*, edited by Kumashiro, M., pp.227-231.

PART VI
Occupational Accidents and Incidents

30. A Study of Work Accidents in Fishery Work

Shuji Hisamune[1], Kiyoshi Amagai[2], Nobuo Kimura[2], Junji Kawasaki[3], Koya Kishida[4]

[1] Maritime Labour Research Institute, Tokyo, Japan
[2] Department of Fishery Science, Hokkaido University, Hakodate, Japan
[3] National Fisheries University, Shimonoseki, Japan
[4] Takasaki City University of Economics, Takasaki, Japan

ABSTRACT

In Japan, the occurrence rate of accidents for fishery workers is eight times as high as that for all industries. Many fishery work accidents are related to poor ergonomics in the working environment. It is particularly important to analyze the various factors causing accidents in fishery work and to apply ergonomics to the conditions on fishing boats.

In this paper, work accidents relating to seine purse and trawler workers were analyzed using fundamental factors (type of fishing, age of worker, work process, cause and injury). After quantification-III analysis, we discerned three factors (cause, work process and age) in fishery work accidents.

The main causes of work accidents were waves for fishery workers under 29 years old, and risky conditions for those from 50 to 59 years old. The pitching and rolling of the ship were found to affect operations, which require skill and experience from the workers. The results bring us to the obvious recommendation to follow safe work procedures and the safety rules for fishery work.

Keywords: *Aging, Fishery work, Work accidents*

INTRODUCTION

The Japanese fishery catch reached the maximum level in the world 20 years ago, but at present it is in four place due mainly to the international fishery situation and declining resources. A large number of fishermen have been drawn to other industries. As a result, the fishing industry suffers from lack of younger workers and the aging of fishery workers. Despite the application of new technologies to fishing boats operations with fewer workers, the number of accidents has not decreased. In Japan, the occurrence rate of accidents for the fishing industry is eight times as high as that for all industries [1]. It is necessary to study the factors relating to work accidents in fishery work, especially purse seine and trawler fishery. The number of worker accidents in purse seine fishery in 1998 was 290, accounting for 23.8% of the 1,221 accidents in the industry as a whole. The number of worker accidents in trawler fishery in 1998 was 289, accounting for 23.4% of the 1,221 accidents. The total number of worker accidents in both sectors in 1998 was 579, accounting for 47.4% of the 1,221 accidents in the industry. The number of fatalities in the purse seine sector in 1998 was 10, accounting for 17.2% of 58 fatalities in the industry. The number of fatalities in the trawler sector in 1998 was 12, accounting for 20.1% of the 58 fatalities in the industry. The total for the two sectors was 22 fatalities, accounting for 38.0% of the 58 fatalities in the industry. Typical work accidents result from workers falling, falling into the sea and being caught in machines. It is particularly important to examine the ergonomic conditions of fishing boats, as they seem to relate to various factors causing accidents in purse seine and trawler fishery.

Purse Seine Fishery: In purse seine fishery, the netting encircles the fish and aquatic animals, hampers their escape and gradually tightens until they are finally bagged. The purse seine is operated by one light ship, two gathering ships and one surrounding net ship. Work accidents are frequent on these ships because of the complex operations essential tosurrounding-net fishing [2].

Trawler Fishery: This type of fishery operates using a trawl net (comprising a cod end, two wing nets extending from both sides of its opening, and two lines) attached to the bottom of the sea and towed by vessels [2].

The guidebook for work accident prevention published by the Fishery Agency and the work procedures published in Shimane Prefecture point out the dangers of surrounding net fishery [3]. A study of energy consumption at work in the fishing industry done by the Maritime Labour Research Institute also reevaluated strenuous type of works [4]. The motion study for purse seine fishery thus seems important. Kawasaki investigated the affect on the mental load of the crew moving over the hull in long-range sailing [5]. With regard to fishing operations, the factor of each of the hulls and the human bodies is a compound one; analysis is difficult and there are few research examples [6]. K. Mikami analyzed the fishing position, i.e., the operating posture of the fishermen, for surf clam fishery, including accident-prevention measures, wotj regard to winch operation [7]. M. Torner investigated the questionnaire of subjective of the fishermen in Sweden analyzing a tendency according to the fish species, the type of job, and the age. It was proposing the ergonomic improvement to have considered a low cost and reducing the workload [8,9]. This research analyzed an offing purse seine fishery in the operation based on the conventional research result.

The modernization of fishing by Fishery Agency took account of the stability of the wage and the securing of work safety, reducing labor by the mechanization of the fishery without study of the movement of the fishermen. As a result, the increase of the operational equipment on deck impeded the hull stability [10-14] . Therefore, to improve fishery work, the operation and the movement of the fishermen must be studied. At present due primarily to a decrease in the number of fishermen, it is difficult to train them during work in a point of strategy manpower and time to spare.

In this study, we focused on various work accident according to the age. The result will be used as a basis for self-checking of these operation and for developing relevant training materials.

METHODS

We analyzed the data of the work accidents about the seine purse and trawler workers, which were reported to Ministry of Transport of Japan in 1998.

The work accidents were divided into subgroups according to the fundamental factors (type of fishing, work process, cause and injured parts).

We analyzed the frequencies of fundamental factors in subgroup according to the age.

The statistical method used in this study was the quantification-III analysis (Hayashi, 1966) based on Factor analysis. We discerned three factors (cause, work process and age) in the accidents during fishery work.

Factor analysis.
While components are liner combinations of the observed variables, factors are liner combinations of unloading variable. The usual factor analysis models express each variable as a function of factors common several variable and factor unique to variable.

$$Z_j = a_{j1}F_1 + a_{j2}F_2 + \ldots + a_{jm}F_m + U_j$$

Where:
   z_j=the jth standard variable
   F_i=the common factors
   m=the number of factors *common* to all the variables
   U_j=the factor unique to variable z_j
   A_{ji}=the factor loadings

Ideally, the number of factors, *m*, will be small, and the contribution of the unique factors will also small. The individual factor loadings, a_{ji}, for each variable should be either very large or very small so each so variable are associated with a minimal number of factors. Thus, you want to explain the observed correlations using as few factors as possible. The unique factors, U's are assumed to uncorrelated with each other and with the common factors [15].

The work accidents about the seine purse workers were divided into subgroups according to the fundamental factors (injured parts, type of fishing, age, work process and cause). The injured parts divided into subgroups according to the head, the body, the arm and the foot. The works divided into subgroups according to operating, unloading, maintenance and sailing. The causes divided into subgroups according to the judgment, behavior, machine and environment.

When comparing the frequencies of fundamental factors in subgroup to according to age, the work accidents about the seine purse and trawler workers were analyzed by the fundamental factors (type of fishing, age, work process, cause and injured parts). After quantification-III analysis, we discerned three factors (cause, work process and age) in the accidents during fishery work.

RESULTS

The work accidents about the seine purse workers

The work accidents about the seine purse workers were divided into subgroups according to the fundamental factors (injured parts, of fishing, age, work process and cause). The injured parts divided into subgroups according to the head, the body, the arm and the foot. The works divided into subgroups according to operating, unloading, maintenance and sailing. The causes divided into subgroups according to the judgment, behavior, machine and environment (Figure 30.1).

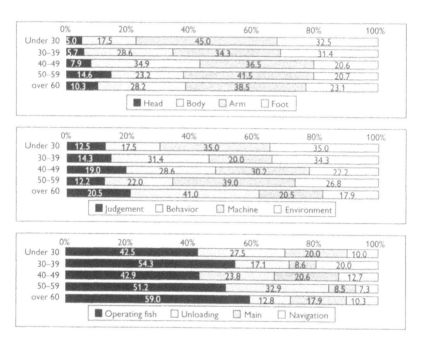

Figure 30.1 The work accidents about the seine purse workers

The occurrence of work accidents was the highest in those who were 50-54 years old.

The fishermen were divided according to age, the injured parts in work accidents, among those who were 20-29 years old and younger, were the arm and the leg, among those who were 50 years old and older, were the head and the body.

The main work process in work accidents, among those who were 30-39 and 60 years old and older, were the operating of fishing, among those who were 40-49 years old, was the unloading.

The main cause in work accidents, among those who were 30-39 years old and younger, was the environment because they could not adapt enough to the work on the sea. The main cause in work accidents, among those who were 50-59 years old, were the machines and among those who were 60 years old and older, were the behaviors.

The work accidents about the trawler workers

The work accidents about the trawler workers were divided into subgroups according to the fundamental factors (injured parts, type of fishing, age, work process and cause) (Figure 30.2).

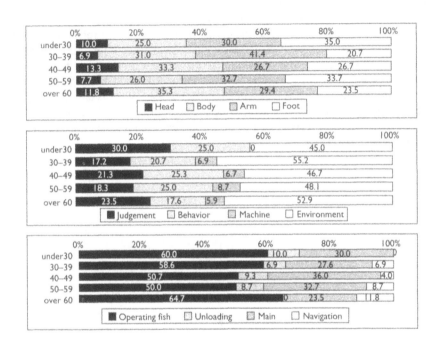

Figure 30.2 The work accidents about the trawler workers

The occurrence of work accidents was in highest in those who were 50-54 years old.

The injured parts in work accidents, among those who were 50-59 years old, and 30-39 years old and younger, were the arm and the leg and among those who were 40-49 years old and 60 years old and older, were the head and the body. The main work process in work accidents, among those who were 30-39 years old and younger, and 60 years old and older, was the operating of fishing and among those who were 40-59 years old, was the maintenance.

The main course in work accidents, among those who were 30-39 years old, and 60 years and older, were the environment. Among those who were 20-29 years old and younger, were the judgment, they could not enough to adapt the work on the sea too.

QUANTIFICATION-III ANALYSIS

The work accidents about the seine purse workers

The work accidents about the seine purse workers were analyzed by the fundamental factors (age, work process, cause and injured parts). The items concerned with work accident were classified 3 factors: cause (factor I), work process (factor II), injured parts (factor III). After quantification-III analysis, we discerned three factors (cause, work process and age) in the accidents during fishery work. After rotation, the first factor account for 38.25 % of the variance, the second account for 24.5 % and the third account for 22.8%. The first three factors account for 85.4% of the total of variance. Figure 30.3 shows the plot of the first factor scores and second factor scores and outline near the age of the plot. Among fishermen were 30 years old and younger, they were close to the effect of waves, the operating of fishing net, the arm and elbow. Among those were 30 years old and younger they have injured their arm and elbow due to the effect of waves.

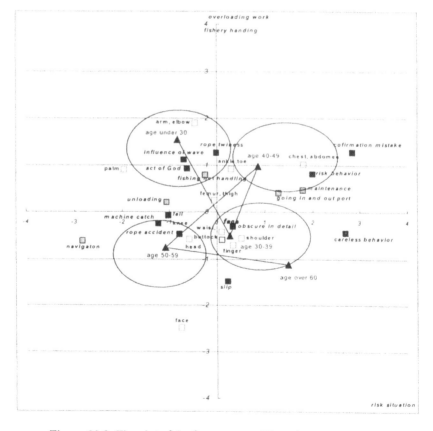

Figure 30.3 The plot of the factor scores (The seine purse workers)

Among fishermen were 30-39 years old they were close to the error behavior, the fishing operation, the shoulder, the finger and the hip, which means they have injured their shoulder, the finger and the hip due to the error behavior.

Among fishermen were 40-49 years old they were close to the risk behavior, the fishing operation, the foot, which means they have injured their foot for the risk behavior.

Among fishermen were 50-59 years old they were close to the clash the machine, the risk environment, the unloading, the head and the ankle, which mean they injured their head and ankle due to the clash with the machine and the risk environment.

Comparing the subgroup according to the age, it was clear that among fishermen who were 20-39 years old, the main cause were the effect of the wave and the error behavior, and among those who were 50-59 years were the risk environment.

The sequences safety and necessary of careful operation in poor environment of catches should be taken into account.

The work accidents about the trawler workers

The work accidents about the trawler workers were analyzed by the fundamental factors (type of fishing, age, work process, cause and injured parts). The items concerned with work accident were classified 3 factors: cause (factor I), work process (factor II), injured parts (factor III). After rotation, the first factor account for 37.2 % of the variance, the second account for 34.6 % and the third account for 18.8%. The first three factors account for 90.3% of the total of variance. Figure 30.4 shows the plot the first factor scores and second factor scores and outline near the age of the plot.

Among fishermen were 30 years old and younger they were close to the unloading work, the head and should, which mean they injured their head and should for the unloading work. Among fishermen were 30-49 years old they were not close to the factor, which mean they have injured various factors.

Among fishermen were 50-59 years old they were close to the maintenance, which mean they injured their hip in the maintenance. Among fishermen were 60 years old and older they were close to the arm and elbow.

Comparing the cause subgroup according to the injured parts and work process, the effect of wave, slipping, the error behavior, the unloading and feet were close. The fishermen have injured their foot in unloading due to the effect of wave, slipping and the error behavior. It needs to improve the equipment and the methods of fishing work.

Comparing work accidents about the seine purse and trawler workers

The work accidents about the seine purse and trawler workers were analyzed by the fundamental factors (type of fishing, age, work process, cause and injured parts). The items concerned with work accidents were classified 3 factors: type of fishing (factor I), cause (factor II), injured parts (factor III). After the quantification-III analysis, we discerned three factors (type of fishing, cause and work process) in the accidents during fishery work.

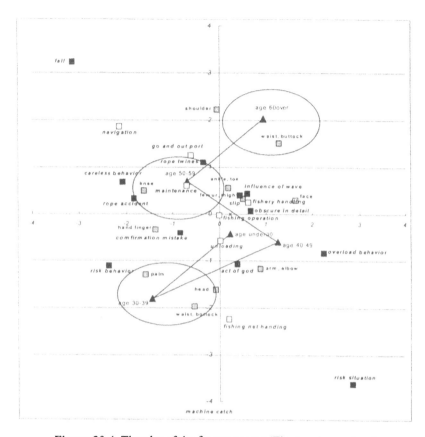

Figure 30.4 The plot of the factors scores (The trawler workers)

The first factor account for 45.6 % of the variance, the second account for 13.7% and the third account for 11.8%. The first three factors account for 71.2% of the total of variance. The first factor was high level. Figure 30.5 shows the plot the first factor scores and second factor scores. We analyzed the fundamental factors in subgroup according to the type of fishing. At the left and right side of this plot, (the positive and negative) loading for factor 1 fall into two distinct clump. The correlations of variables in the right group (seine purse) were close to the clashing with the machines, the fall, the navigation, the unloading, the head, the hand, and the finger and the shoulder. The fishermen for seine purse workers have injured their head and shoulder and the hand due to the clashing machine and the fall. The correlations of variables in the left group (trawler), were close to the operating the net, the risk behavior, the effect of wave, the head, the ankle and the bless. The fishermen for workers have injured their head, the ankle and the bless due to the effect of wave. Variable near the intersection (the point 0,0) were associated the operating fishing, the leg and the foot. These were common factor for both type of fishing.

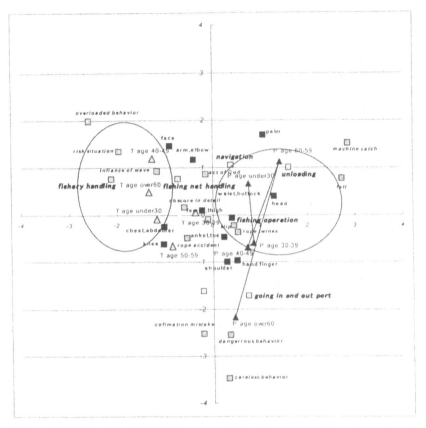

Figure 30.5 The plot of the factor scores (The seine purse and trawler workers)

DISCUSSION

The work accidents about the seine purse and trawler workers were analyzed by the fundamental factors (type of fishing, age, work process, cause and injured parts). After quantification-III analysis, we discerned three factors (cause, work process and age) in the accidents during fishery work. The main cause of work accidents under 29 years old workers was the wave, and the main causes of work accidents from 50 to 59 years old workers was the risk condition. It is important to study the aging and the types of fishing, and organize the operations in a safe manner.

The pitching and rolling of the ship were found to affecting the operations. Many of fishery work accidents are related to poor ergonomics condition of the working environment. It seems important to improve the fishing machines and equipment. Special care was obviously needed to follow safe work procedures. The needs to apply the sequences of safety and careful operations in poor environment of catches should be taken into account. The fishermen should check the work by themselves and propose improved methods of fishery. Such a manual can be used in fishing technology education so as to show the appropriate flow of work. This work manual can be provided fishery companies through the Association for Promoting Safety and Sanitation for Seafarers.

This research was the 7th seamen's accidents prevention plan in Ministry of Transport of Sea fares Department, Maritime Technology and Safety Bureau. We appreciated the cooperation Ministry of Transport and Association for Promoting Safety and Sanitation for Seafarers with valuable data.

REFERENCES

1. The Ministry of Transport sea technology safe bureau seafarers dept. In *The book of the amount of the sailor disaster disease occurrence situation report* (the article of 111th of the sailor low), 1997, P6 and P37.
2. Kaneda, Y., 1995, *Fisheries and fishing methods of Japan*, (seizando).
3. *The problem, the fishing boat to discuss in the reduce labor seminar*, 162, 1969, (The fisheries agency), pp. 30-49.
4. The maritime labor science laboratory. In *The research about the actual state of the labor of firewood knitting the fishermen*, 1970.
5. Kawasaki, J., Amagai, K., Kimura, N. and Hokimoto, T., 1992, The Response Amplitude Characteristics of Human Responses to the Ship Motions. In *The journal of Japan Institute of Navigation*, 87, pp. 79-88.
6. Kawasaki, J., Amagai, K. and Kimura, N., 1995, A Study for Human Factors on the Occasion of the Collision of Fishing Vessels Caused by Insufficient Lookout in the Coast of Hokkaido. In *The journal of Japan Institute of Navigation*, 93.
7. Mikami, K., 1989, *The doctor's theses*, Hokkaido University.
8. Torner, M., Blide, G., Eriksson, H., Kadefors, R., Karlsson, R., Peterson, I., and Musculo-skeletal, symptoms as related to working conditions among Swedish professional fishermen. In *Applied ergonomics*,19.3, pp. 191-201.
9. Torner, M., Blide, G., Eriksson, H., Kadefors, R., Karlsson, R. and Peterson, I., 1998, Workload and ergonomics measures in Swedish professional fishing. In *Applied Ergonomics*,19.3, pp. 202-212.
10. Hamuro, K., 1970, As for the reduce labor model design of the purse seine fishing boat. In *The fishing boat*, 169, pp. 20-33.
11. The fishing modernization seminar: The recent fishing gear lecture. In *The fishing boat*, 187, pp. 53-63, 1973.
12. Kayahira, M., 1974, As for the excessive equipment with the purse seine fishing boat. In *The fishing boat*, 194, pp. 71-78.
13. Hamuro, K., 1977, The overview, the fishing boat of the research result of the rationalization seminar of the purse seine. In *The fishing boat*, 212, pp. 55-64.
14. Matsubayashi, N., 1992, As for the reduce labor system of the purse seine fishing ship. In *The fishing boat*, 298, pp. 89-93.
15. SPSS Base 7.5 Applications Guide, 1997.(0).

31. Renal Function Decline in Aged Workers Enhances Toxic Effect of Occupational Chemicals

Kan Usuda, Koichi Kono, Takemasa Watanabe, Tomotaro Dote, Hiroyasu Shimizu, Chisato Koizumi, Mika Tominaga, Mitsuya Akashi

Department of Hygiene and Public health, Osaka Medical College, Osaka, Japan

ABSTRACT

It is estimated that more than 200,000 patients are treated by dialysis in Japan, and 20% of them have a more than 10-year history of dialysis. In addition to the increase in the number of patients under dialysis over these decades, a marked increase in the mean age of patients is observed. Currently in Japan, 20% of patients under dialysis over 60 years of age are employed.

As the toxicity caused by industrial chemicals is enhanced by aging factors such as decreased organ function, recognizing the toxicokinetics and toxicodynamics of industrial chemicals relative to aged workers is important. In this respect, aged workers under dialysis may be a useful model case to assess fitness to work and evaluate the risk of hazardous chemicals.

The focus of this study is on the variability of serum ionic fluoride (F^-) in 29 dialysis patients. Fluoride compounds are often used in etching and cleaning operations, and solvents are used in photoresist at semiconductor manufacturing facilities. Due to the recent development of a high-technology-society, fluoride demand and consumption have been increasing year by year.

Although the result confirmed notable dializability of F^-, serum F^- of the patients before and after dialysis was statistically higher than the controls, and a tendency of F^- accumulation was suggested. This conclusion could suggest the establishment of guidelines for the treatment of workers with aging organ problems when they handle toxic chemicals.

Keywords: *Aging, Aged worker, Fluoride, Hemodialysis*

INTRODUCTION

As the population of aged people increases, the number of industrial workers who have chronic organ failure or functional disorders also grows. Developments in medical care over the past decades have provided remarkable life-sustaining technologies that have contributed to prolongation of patient life span, as well as to a general increase in health and well-being. End-stage renal disease treatment is a prototype of such technologies [1].

It is a fact that in Japan, treatment of end-stage renal disease depends disproportionately heavily on hemodialysis and almost negligibly on renal transplants from cadaveric or living donors. There are uniquely Japanese attitudes toward and conditions for organ transplants [2]. Today, Japan has the world's largest population of dialysis patients. Nearly 170,000 patients were treated by dialysis in Japan in 1996 [3]. Today the number is estimated to be more than 200,000, and owing to progress and improvement in dialysis modalities, it is estimated that approximately 20% of them have a more than 10-year history. Thus, this disease represents a major problem in the elderly, and the number of dialysis patients will indeed increase as the aged population increases. In addition to the increase in dialysis patients, a marked increase in the mean age of patients is observed; it has exceeded age 60. Currently in Japan, 20% of dialysis patients over 60 years old are employed.

As the toxicity caused by chemicals is enhanced by aging factors such as decreased organ function, aged workers under dialysis may be a useful model case to assess fitness to work and evaluate the risk of hazardous chemicals. Patients undergoing long-term dialysis are known for ailments relating to metal elements, such as dialysis aluminum encephalopathy [4] and osteodystrophy [5], deficiency of zinc [6], magnesium [7] and other trace elements, or excess of several contamination elements [8]. Fluoride (F) is one of these elements, and abnormal high profiles in dialysis patients are reported [9,10,11].

F is a naturally occurring ubiquitous element. F, the 13th most abundant element in the earth's crust, is never encountered in its free state in nature. It exists only in combination with other elements as F compounds. It is found in this form as a constituent of minerals in rocks and soil everywhere. Water passes over rock formations containing F and dissolves these compounds, creating ionic fluoride (F^-). The result is that small amounts of soluble F^- are present in all water sources, including the oceans [12,13,14]. Although F is used for the fluoridation of domestic city water and for toothpaste to prevent dental caries in many countries, adverse effects are also discussed [15]. F is found also in our daily food, especially in Japan, where the populace is fond of F-rich food and the daily amount of F intake is comparatively larger than that in other countries [16]. F is known to be easily, rapidly and completely absorbed from the gastrointestinal tract and absorbed F is nearly all excreted via the kidney within 24 hours after ingestion, and the discharge of F into the urine depends on renal clearance [17]. In the environmental field, endemic fluorosis is a cause of osteosclerosis, mottled teeth and stained teeth [18]. It also has an effect on the metabolism of other trace elements. These are serious public health problems in many developing countries, where drinking water with a high F content is found. In industry, F compounds are often used in etching and cleaning operations, and solvents are used in photoresist at semiconductor manufacturing facilities [19,20]. Due to the recent development of a high-technology-society, F demand and consumption have been increasing year by year.

In this study, we studied the movement of ionic F^- across the dialysis membrane in dialysis patients. The aim of this study is to assess the disturbance of F^- in dialysis patients as the model of aged organ function decline, and to compare the clearance of

F⁻ by dialysis with that of uremic elements such as creatinine (Cr), blood urea nitrogen (BUN) and phosphorus (P).

MATERIALS AND METHODS

To investigate the movement of F⁻ across the dialysis membrane, serum and dialysate F⁻ of 29 dialysis patients were determined. They live in an urban area and are free from environmental or occupational F exposure. The group was composed of 10 males and 19 females, mean age 58.9 ± 10.6 years, with a range from 41 to 83 years. They all had undergone 4h of hemodialysis 3 times per week in a hospital. The average duration of dialysis was 7.4 ± 6.3 years, with a range from 1 to 23 years. They were mainly under dialysis using a single-use hollow-fiber type dialyzer. The primary renal diseases of the patients were chronic glomeronephritis, diabetic nephropathy and polycystic kidney.

Regarding diet, guidelines were given about controlling the intake of protein.

The serum samples were collected before and after dialysis from the inlet tube of dialyzer at the beginning of dialysis and outlet tube of dialyzer at the end of dialysis. The dialysate samples were collected from the in and outlet tubes of the dialyzers at the beginning of dialysis.

The dialysate was prepared by diluting Solution A (ionic adjustment) and Solution B (sodium bicarbonate solution for pH adjustment) with RO water (reverse osmosis water) in the ratio of 1: 1: 34. Solution A, Solution B and RO water were also sampled.

Serum Cr, BUN and P were also investigated for a total of 92 dialysis patients, which included the above mentioned 29 subjects and who had similar conditions.

F⁻ was determined by an Orion Research Model EA940 digital ion analyzer and its fluoride selective electrode [21].

The dialysis clearance in the subjects was calculated according to the following formula [22].

Dialysis clearance (ml/min) = $[Q_{BO} (C_{BI} - C_{BO})/C_{BI}] + Q_F$,
where Q_{BO} = blood outlet flow rate from the dialyzer (ml/min);
C_{BI} = serum concentration of the elements on the inlet side of the dialyzer at the beginning of dialysis;
C_{BO} = serum concentration of the elements on the outlet side of the dialyzer at the end of dialysis;
Q_F = ultrafiltration volume rate (ml/min)

The mean Q_{BO} rate of 200 ml/min and Q_F rate of 10 ml/min were applied to the formula.

The differences between the means were evaluated by using the paired or unpaired t test.

RESULTS

Table 31.1 shows the mean serum F⁻ of the patients before and after dialysis [11], F exposed workers [21], and non-F exposed control subjects [23]. Serum F⁻ of the patients before and after dialysis and post shift level of F⁻ workers was statistically higher than those of control subjects (95%, p<0.001).

Figure 31.1 shows the change of serum F⁻ during dialysis [11]. As shown in this figure, F⁻ decreased with a statistically significant difference (95%, p<0.001); however, it failed to return to normal.

Figure 31.2 shows the change of F⁻ in the dialysate during dialysis [11]. Outlet side F⁻ statistically increased compared with inlet side (95%, p<0.01).

Table 31.2 shows the mean F⁻ of each dialysate [11]. In these dialysis component solutions, solution B had an exceptionally high F⁻. F⁻ in solution A and RO water was far lower than that in solution B. Finally, high F⁻ in solution B was diluted mainly with RO water. Therefore, F⁻ is sufficiently lower in the mixed solution than in serum before dialysis.

Table 31.3 shows the mean serum Cr, BUN and P before and after dialysis in 92 patients [11]. As shown in this table, serum Cr, BUN and P decreased remarkably after dialysis.

Table 31.4 shows the mean dialysis clearance of fluoride, Cr, BUN and P [10,11]. As shown in this table, the dialysis clearance of F⁻ was statistically significantly lower than those of Cr, BUN (95%, p<0.001) and P (95%, p<0.01).

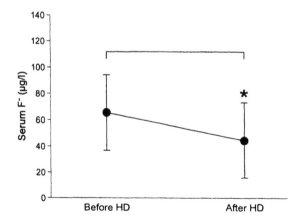

Figure 31.1 Change in serum F⁻ during dialysis (n=29).
Data represent the mean SD. Vertical bars denote SD.
*p<0.001: significant difference in comparison to before dialysis [11]

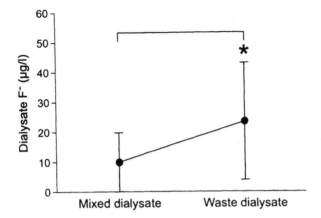

Figure 31.2 F⁻ in the mixed and waste dialysate (n=29). Data represent the same as Figure 31.1. *p<0.01: significant difference from the mixed dialysate [11]

Table 31.1 Serum F⁻ (means±SD) in patients before and after dialysis [11], healthy controls [21] and hydrofluoric acid (HF) worker [23]

Subjects	n		F⁻, μg/l
Patients before HD	29		65.9±28.3*
Patients after HD	29		46.6±27.0*
HF worker	19	preshift	19.9±1.5
		postshift	46.8±17.1*
Healthy controls	2,656		19.0±8.0

n = Number of subjects.
* p<0.001: significant difference from healthy controls.

Table 31.2 F⁻ (means±SD) of each dialysate used in the dialysis system (n=29) [11]

	F⁻, μg/l
Solution A	11.2±9.16
Solution B	163.2±56.1
RO water	4.45±5.08
Mixed dialysate	7.16±8.91
Waste dialysate	28.7±18.1

Table 31.3 Serum Cr, BUN and P (mg/l; means±SD) in patients
before and after dialysis (n=29) [11]

	Before	After
Cr	11.1±2.63	5.23±1.27*
BUN	83.0±20.6	32.4±10.5*
P	5.99±1.79	3.48±0.757*

* p< 0.001: significant difference from before HD.

Table 31.4 Dialysis clearance (means±SD) of F⁻, Cr, BUN and P [11]

	N	Clearnce, ml/min
F⁻	29	55.6±101.3
Cr	92	113.9±19.3**
BUN	92	129.9±25.7**
P	92	87.2±32.6*

n = Number of subjects.
* P<0.01, ** p<0.001: significant difference from F⁻.

DISCUSSION

Figures 31.1 and 31.2 show a notable dialyzability of F⁻; however, serum F⁻ was not sufficiently reduced and remained high (Table 31.1). It is reported that using low F⁻ dialysate can reduce the high serum F⁻ in dialysis patients [24,25], but serum F⁻ remained high after dialysis in this study. Although trace contamination of F- in solution A and B was confirmed, F⁻ in RO water was sufficiently low for the preparation of the mixed dialysate, and mixed dialysate F⁻ was adequate for the dialysis session (Table 2). When unpurified high F⁻ dialysate was used, F⁻ was not eliminated by dialysis and the patients were exposed to F⁻. In this study, since F⁻ in the mixed dialysate was assumed to be sufficiently lower than in serum before dialysis, it was omitted. Dialysis is a diffusion-driven and size-discriminatory process for the clearance of small solutes, such as Cr, BUN and P. F is known for its low weight. Since the pore size of dialysis membrane filtration holes is larger than the ionic radius of F⁻, dialysis clearance of F⁻ should be higher than that of Cr, BUN and P. So, dialysis clearance of F⁻ was calculated for comparison with other chemicals.

The obtained low clearance of fluoride suggests the existence of combined F. The combined fluoride will prevent the fluoride from passing through the dialysis membrane filtration hole. Many reports say that a fraction of protein-bound element or drug, being unavailable for free diffusion across the dialysis membrane, lowers the clearance or exhibits reduced dialysis [26]. A similar phenomenon may occur with F⁻ [27]. Although the result confirmed notable dializability of F⁻, serum F⁻ of the patients before and after dialysis was statistically higher than the controls, and a tendency of F^- accumulation was suggested.

When patients with chronic renal failure under dialysis are employed at F-exposed work places or live in regions with fluoridated water or polluted areas, they ingest excessive F from their industrial or community environment in addition to their daily intake of food containing fluoride. Japanese foods contain a particularly high proportion of F. An excessive ingestion of F and insufficient serum F purification in patients with dialysis will enhance hard tissue mineral disturbance by F storage. It is shown that, despite the maximum F elimination ability of dialysate and a net clearance of F during a given dialysis procedure, serum F^- failed to return to normal. To reduce serum F^-, patients need to practice dietary control for the restriction of oral F intake. Namely, they should avoid F-rich foodstuffs such as tea and marine products and sea plants. Moreover, they should be monitored for gastrointestinal functions such as gastric acidity, which depends on F absorption, or for endocrine function such as blood gastrine, which is significantly affected by dialysis. Once the kidney is impaired, the dialysis system by itself cannot reduce serum F^- to normal, even if a sufficiently low F^- dialysate is used.

In conclusion, this study emphasizes the importance of controlling excessive intake of F in addition to the importance of an effective and safe F-reducing dialysis system in preventing F retention and bone diseases in dialysis patients. Frequent monitoring of serum F^- in dialysis patients is required, especially if they are engaged in F-exposed work or living in an F-polluted area. Our study could suggest the establishment of guidelines for the treatment of aged workers suffering from decreased organ function.

REFERENCES

1. Cummings, NB., 1990, Ethical issues in geriatric nephrology: overview. In *Am J Kidney Dis* **16**, pp. 367-371.
2. Ohi, G., Hasegawa, T., Kumano, H., Kai, I., Takenaga, N., Taguchi, Y., Saito, H. and Ino, T., 1986, Why are cadaveric renal transplants so hard to find in Japan? An analysis of economic and attitudinal aspects. In *Health Policy* **6**, pp. 269-278.
3. Shinzato, T., Nakai, S., Akiba, T., Yamagami, S., Yamazaki, C., Kitaoka, T., Kubo, K., Maeda, K. and Morii, H., 1999, Report of the annual statistical survey of the Japanese Society for Dialysis Therapy in 1996. In *Kidney Int.* **55**, pp. 700-712.
4. Berend, K., van der Voet, G. and Boer, WH., 2001, Acute aluminum encephalopathy in a dialysis center caused by a cement mortar water distribution pipe. In *Kidney Int.* **59**, pp. 746-753.
5. Lefebvre, A., de Vernejoul, MC., Gueris, J., Goldfarb, B., Graulet, AM. and Morieux, C., 1989, Optimal correction of acidosis changes progression of dialysis osteodystrophy. In *Kidney Int.* **36**, pp. 1112-1118.
6. Erten, Y., Kayatas, M., Sezer, S., Ozdemir, FN., Ozyigit, PF., Turan, M., Haberal, A., Guz, G., Kaya, S. and Bilgin, N., 1998, Zinc deficiency: prevalence and causes in hemodialysis patients and effect on cellular immune response. In *Transplant Proc.* **30**, pp. 850-851.

7. Markell, MS., Altura, BT., Sarn, Y., Delano, BG., Ifudu, O., Friedman, EA. and Altura, BM., 1993, Deficiency of serum ionized magnesium in patients receiving hemodialysis or peritoneal dialysis. In *ASAIO J.* **39**, pp. M801-804.
8. Zima, T., Tesar, V., Mestek, O. and Nemecek, K., 1999, Trace elements in end-stage renal disease. 2. In Clinical implication of trace elements. *Blood Purif.* **17**, pp. 187-198.
9. Usuda, K., Kono, K., Orita, Y., Tanimura, Y., Sumi, Y., Yoshida, Y., Shimahara, M. and Senda, J., 1996, Complex effects of hemodialysis on the metabolism of fluoride. In *ACES* **8**, pp. 91-95.
10. Usuda, K., Kono, K. and Yoshida, Y., 1996, Clearance of fluoride by hemodialysis. In *Clin. Nephrol.* **45**, pp. 363-364.
11. Usuda, K., Kono, K. and Yoshida, Y., 1997, The effect of hemodialysis upon serum level of fluoride. In *Nephron* **75**, pp. 175-178.
12. Tsutsui, A., Yagi, M. and Horowitz, AM., 2000, The prevalence of dental caries and fluorosis in Japanese communities with up to 1.4 ppm of naturally occurring fluoride. In *J Public Health Dent.* **60**, pp. 147-153.
13. Yang, CY., Cheng, MF., Tsai, SS. and Hung, CF., 2000, Fluoride in drinking water and cancer mortality in Taiwan. In *Environ Res.* **82**, pp. 189-193.
14. Hastreiter, RJ., Leppink, HB., Sundberg, LB., Knaeble, DJ., Turtle, DR., Falken, MC. and Roesch, MH., 1992, Clinical implications of an investigation into the occurrence and distribution of naturally occurring fluoride. In *Northwest Dent.* **71**, pp. 19-23.
15. Mascarenhas, AK. and Burt, BA., 1998, Fluorosis risk from early exposure to fluoride toothpaste. In *Community Dent Oral Epidemiol.* **26**, pp. 241-248.
16. Toyota, S., 1979, Fluorine content in the urine and in the serum of hydrofluoric acid workers as an index of health administration Sangyo Igaku. In 1979 21, pp. 335-348.
17. Kono, K., 1994, Health effects of fluorine and its compounds Nippon Eiseigaku Zasshi. 49, pp. 852-860.
18. Wang, Y., Yin, Y., Gilula, LA. and Wilson, AJ., 1994, Endemic fluorosis of the skeleton: radiographic features in 127 patients. In *AJR Am J Roentgenol.* **162**, pp. 93-98.
19. Woskie, SR., Hammond, SK., Hines, CJ., Hallock, MF., Kenyon, E. and Schenker, MB., 2000, Personal fluoride and solvent exposures, and their determinants, in semiconductor manufacturing. In *Appl Occup Environ Hyg.* **15**, pp. 354-361.
20. Swan, SH., Beaumont, JJ., Hammond, SK., VonBehren, J., Green, RS., Hallock, MF., Woskie, SR., Hines, CJ. and Schenker, MB., 1995, Historical cohort study of spontaneous abortion among fabrication workers in the Semiconductor Health Study: agent-level analysis. In *Am J Ind Med.* **28**, pp. 751-769.
21. Yoshida, Y., Toyota, S., Kono, K., Watanabe, M., Iwasaki, K. and Kato, I., 1978, Fluoride ion levels in the biological fluids of electronic industrial workers. In *Bull Osaka Med Sch.* **24**, pp. 56-67.
22. Henderson, LW., 1983, Biophysics of ultrafiltration and hemofitration. In *Replacement of Renal Function by Dialysis*, edited by Drukker, W., Parsons, FM. and Jaher, JF., (Boston, Nijhoff), pp. 252-253.

23. Kono, K., Yoshida, Y., Watanabe, M., Tanioka, Y., Orita, Y., Dote, T., Bessho, Y., Takahashi., and Yoshida, J. and Sumi, Y., 1992, Serum fluoride as an indicator of occupational hydrofluoric acid exposure. In *Int Arch Occup Environ Health*, **64**, pp. 343-346.
24. Canturk, NZ., Undar, L., Ozbilum, B., Canturk, M. and Yalin, R., 1992, The influence of hemodialysis on plasma fluoride. In *Mater Med Pol.* **24**, pp. 89-90.
25. Chaleil, D., Simon, P., Tessier, B., Cartier, F. and Allain, P., 1986, Blood plasma fluoride in haemodialysed patients. In *Clin Chim Acta.* **156**, pp. 105-108.
26. Gwilt, PR. and Perrier, D., 1978, Plasma protein binding and distribution characteristics of drugs as indices of their hemodialyzability. In *Clin Pharmacol Ther.* *24*, pp. 154-161.
27. Milhaud, G., Diagbouga, PS. and Joseph-Enriquez, B., 1992, Fluoride bound to plasma constituents in cattle. In *Fluoride* **25**, pp. 85-91.

32. Characteristics and Perspectives of Occupational Accidents Involving Aged Workers in Korea

Hyeon-Kyo Lim[1] and Masaharu Kumashiro[2]

[1] Department of Safety Engineering, Chungbuk National University,
Mt.48, Gaeshin-dong, Heungduk-ku, Cheongju, Chungbuk, Korea
[2] Department of Ergonomics, IIES,
University of Occupational and Environmental Health,
1-1 Iseigaoka, Yahatanishi-ku, Kitakyushu, Japan

ABSTRACT

Recently, South Korea has suffered severe setbacks due to its weakened economy. In this research, the impact of these setbacks was reviewed, and the work arrangements and occupational accidents of aged Korean workers over the last five years were analyzed from the aspect of accident frequency, severity, and characteristics.

The results show that most middle-aged and aged workers work in the manufacturing industry, followed by the public/personal service industry. Their employment rates were not much higher than those of the total economic population of the same age.

Meanwhile, as many as 40% of occupational accidents involved middle-aged and aged workers, and the percentage of accidents involving aged workers alone is as high as 25%. Though it is often assumed that accidents would be less frequent but more severe as the worker's age increases, the accidents are more frequent and more severe. It is demonstrated that as years go by, the mean average of workdays lost increases regardless of age group. Major accident types are falling and slipping, followed by work-related diseases, and the proportion of slipping is higher than any other type of accident.

Based on these results, it is concluded that accident characteristics reflect the domestic industrial situation of South Korea, and that the trend for these accidents to increase is likely to continue as the proportion of aged workers rises.

Keywords: *Aging, Occupational accidents, Accident frequency, Accident severity, Accident characteristics*

INTRODUCTION

According to a report published by the National Statistical Office of Korea, the population of South Korea was estimated to be 46.13 million as of November 1st, 2000. This means that the population has risen 1.5 times from the 32.2 million total of

1970, and that South Korea ranks 26th among nations in the world in terms of population size.

Meanwhile, as in other developed countries, the birth rate, death rate, and population growth rate are declining. The mean life expectancy at birth already exceeded 70 years of age in 1997, and the mean age of the population continues to rise. In consequence, the aged population over 40 has kept on increasing as shown Fig. 32.1, and the aged older than 50 account for as much as 25% of the economic population. Therefore, the population aged 65 or older is estimated to reach more than 10 million by 2030.

Thereby, the ratio of the elderly population (65 or older) to the young population (14 or younger), or the Index of Aging, stood at the 35.0 level as of November, 2000. This is somewhat lower than that of other OECD countries, though it is projected to reach 50 in 2010, 76.5 in 2020, and finally to rise as high as 120.3 in 2030. It would then be on a similar level of such aged societies as contemporary Italy (127.3), Greece (119.6), or Japan (115.5) (National Statistical Office, 1996).

To summarize, South Korea already has become an aged society, and in particular, aging is proceeding at more rapid pace than other countries. This can be demonstrated by the December 1996 National Statistical Office estimate that the Index of Aging would be 32.89 in 2000, but in fact it already reached 35.0 in 2000 - an error of 10%. Consequently, it can be easily anticipated that lack of manpower in industrial plants will worsen and the proportion of middle-aged and aged workers will continue to grow.

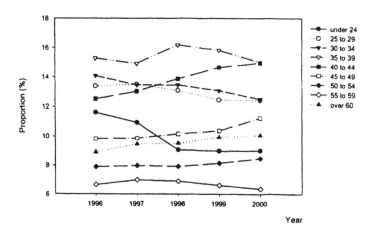

Figure 32.1 Proportion change of age group

WORK ARRANGEMENTS OF AGED PEOPLE

The work arrangements of aged people older than 50 rose continuously from 245.5 million in 1990 to 345.6 million in 2000. Furthermore, the overall unemployment rate of aged people rose to 3.3% in 1998 and to 3.5% in 1999 due to the severe setbacks resulting from the weakened economy following the IMF bailout and corporate restructuring in 1997. The employment market has shown the first signs of resurgence with a gradual recovery in economic activities, however. This resulted in an unemployment rate at the 2.0% level in 2000 (National Statistical Office, 2001).

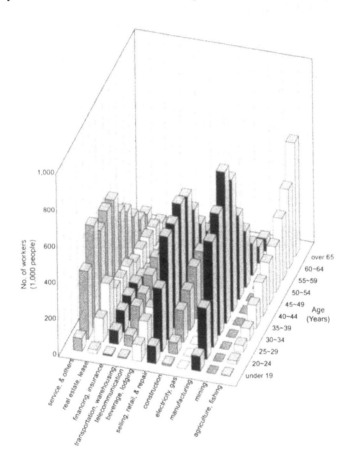

Figure 32.2 Employment of industrial sectors by age groups (2000)

As of 2000, a look at the industrial sectors employing middle-aged and aged workers older than 45 years of age showed that the manufacturing industry had the highest percentage, followed by the public/personal service sector, as noted in Fig. 32.2. Though the number of employed workers was large, it did not necessarily mean that the employment rate of aged workers was high. The employment rate was highest in the public/personal service sector, followed by the financial, insurance, and real estate sectors. In contrast, the manufacturing, wholesale and retail, restaurant, and hotel sectors had relatively low employment rates.

Aged workers were most frequently employed in elementary occupations, followed by crafts, machine operation, assembly, service work, and sales, as shown.

PROPORTION OF OCCUPATIONAL ACCIDENTS INVOLVED MIDDLE-AGED & AGED WORKERS

In recent years, the number of occupational accidents has been steadily declining, while the rate of severity has been rising. It is not unconnected with the continual increase in the ratio of occupational accidents involved middle-aged and aged workers.

As shown in Fig. 32.3, the proportion of occupational accidents involving middle-aged and aged workers older than 45 years of age has been rising recently. Thus, the proportion of occupational accidents involving these age groups in 2000 accounted for as many as 38% of all occupational accidents, and that of aged workers over 50 alone was as high as 25%.

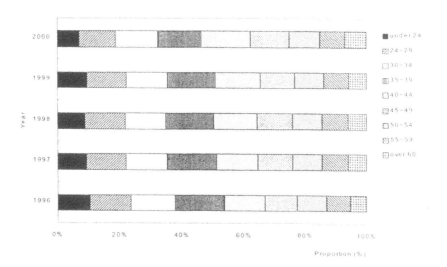

Figure 32.3 Composition of victims in occupational accident types by age

This fact can be easily understood by comparing the economic population and occupational accidents by age group. Fig. 32.4 shows a comparison of ratios between the proportion of occupational accidents and the proportion of economically active populations by age. As shown in the figure, the accident frequency of older workers (40 or over) was relatively higher than the proportion of employed workers and kept rising steadily during last five years. It therefore can be inferred that this trend is likely to continue for years as long as the number of older workers increases.

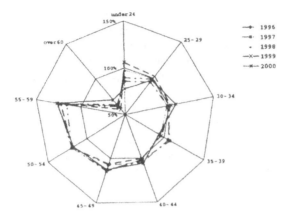

Figure 32.4 Comparison between proportions of accidents and population

CHARACTERISTICS OF OCCUPATIONAL ACCIDENTS INVOLVING AGED WORKERS

Fig. 32.5 shows the mean number of occupational accidents per 1,000 workers by age. It shows that, with the exception of the two groups aged under 18 or over 60, the mean number of occupational accidents increases with age. This implies that the accident frequency increases as age increases regardless of industry. Some previous researchers obtained a different conclusion through their research, however (Simonds and Shafai-Sahrai, 1977; Leigh, 1986). This result is in accordance with other studies (Landen and Hendricks, 1992), and may be considered an indication of conditions in Korean industrial plants. They are not sufficiently developed as an environment for aged workers and are mainly administered by the productive-age population.

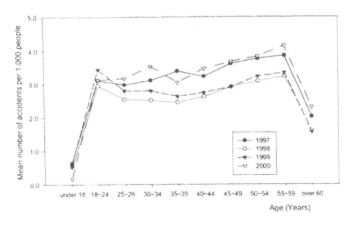

Figure 32.5 Mean number of accidents per 1000 workers

Meanwhile, Fig. 32.6 shows the variation of the mean lost workdays over age. As expected, the mean lost workdays of occupational accidents tend to generally increase, demonstrating that the severity of injury increases as workers get older. This result concurs with previous research (Dillingham, 1981; Landen and Hendricks, 1992).

Yet, it should be stressed that with the passage of time, the mean average of workdays lost increase regardless of age group, though there are some fluctuations. Compared to 1997, the mean average of workdays lost in 2000 increased for all ages by 25% (in the age group of over 60) to 38% (in the age group of 40 to 44). Considering that this shows the variation of averaged data over all industries and there must not have been abrupt changes in work patterns or workloads, it implies that there are apprehensible global changes in work abilities or incident escape abilities for all ages.

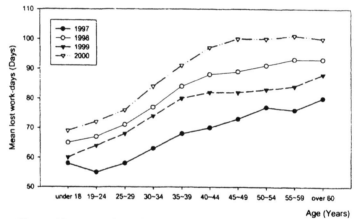

Figure 32.6 Variation of lost workdays of occupational accidents

Accident types occurring in the year of 2000 are summarized by age in Fig. 32.7. As shown, the highest level of most accident types was recorded at roughly 40 years of age. Therefore, the 40-44 age group had the most victims. It indicates that the members of this group play the most important and active role in their work than any other group.

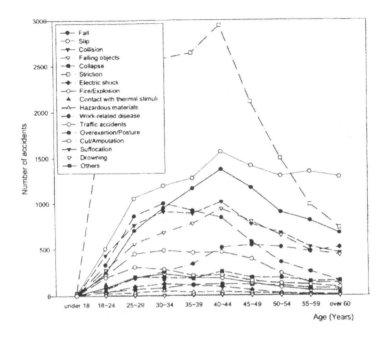

Figure 32.7 Number of occupational accidents by age in Korea (2000)

Work-related disease does not decrease after 40 years of age, but the occurrence rate is maintained thereafter. In addition, slipping is a more frequent accident type than falling for workers older than 50. It implies that because of diminished body resilience with increasing age, especially the maintenance of body balance, such age-related accidents such as slipping and falling, and work-related diseases occur more frequently than other types of accidents, such as constriction, overexertion, awkward posture and the like.

The seriousness of the problem becomes apparent when it is considered that the number of accidents involving highly aged people is much fewer than those involving younger people. Fig. 32.8 compares the percentages of occupational accidents by age group. It shows that falling, slipping, being struck by flying/falling objects, and work-related disease. gradually came to account for a greater percentage of accidents as worker age increased, while the percentage of constriction, overexertion, or awkward postures decline.

Based on these findings, it can be concluded that accidents involving older workers could result from impaired physiological function, while accidents involving younger workers might be explained by inexperience (Laflamme, et al., 1995).

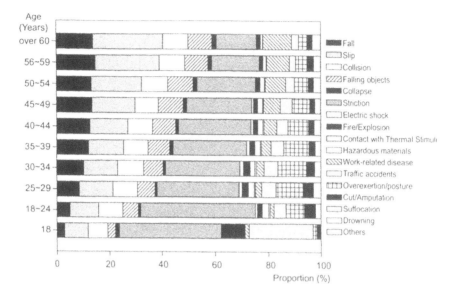

Figure 32.8 Composition of occupational accidents by age (2000)

DISCUSSIONS AND CONCLUSION

To summarize the variation trend of the economic population in South Korea, the economically active population is gradually increasing while the aged population is drastically increasing. The government has not yet provided practical solutions to these social phenomena, however.

According to the 'Promotion Measures for Employment of Aged Workers' compiled by the Ministry of Labor in 1995, the South Korean government plans to advise enterprises to gradually change their policies and lift the retirement age to 60 from the current 55, which is based on enterprise policy or collective agreements. It advises governmental organizations or public corporations to form new contracts different from the old ones with aged workers after their retirement.

To increase the employment rate of aged people in appropriate occupations to 80%, the South Korean government adopted a policy of giving high priority to aged workers when governmental organizations or public corporations employ new workers. The reality is quite different from this policy, however. Though the unemployment rate of aged people older than 55 in the economically active population had been around 0.6% by 1996, it soared to 3.5% in 1999 due to the economic crisis and fell again to 2.0% in 2000, as previously noted. At its highest, however, the employment rate was 6.88 % for enterprises employing more than five workers in 1998, and a much lower 3.51% for those employing more than 300 workers (Ministry of Labor, 1999).

Meanwhile, since 1999, the economically active population declined the most in the group aged 25 to 29, whereas it increased the most in the group aged 60 or older. Thus, the unemployment rate increased only in that group.

Consequently, it is quite natural to imagine that more aged people would desire employment for their living expenditures as well as their work lives, without consideration of their bodily functions or occupations. This assertion is supported by the recent remarkable increase in the number of machine operators, craft workers, and elementary occupations - by 421,000 people in 1999 alone - and the decline in the number of regular employees contrasted with an increase in temporary employees and daily workers.

On the other hand, the manpower shortage rate is said to have increased again recently in all industries due to the business recovery. It is also higher than average in the construction and the manufacturing industry, which is noteworthy.

To conclude, age-related hazards would be one major factor in occupational accidents, and the number of accidents involving aged workers would eventually rise without efforts to prevent age-related occupational accidents and to improve employment policies for the aged population. The development of jobs and worksites for aged workers is an urgent task.

REFERENCES

1. Dillingham, A.E., 1981, Age and workplace injuries. In *Aging and Work*, **4**, pp. 1-10.
2. Laflamme, L. and Menckel, E., 1995, Aging and occupational accidents: thirty years of conflicting findings. In *The Paths to Productive Aging*, edited by Kumashiro, M., pp. 187-193.
3. Landen, D.D. and Hendricks, S.A., 1992, Estimates from the National Health Interview Survey on occupational injury among older workers in the United States. In *Scandinavian Journal of Work, Environment and Health*, **18**, Suppl.2, pp. 18-20.
4. Leigh, J.P., 1986, Individual and job characteristics as predictors of industrial accidents, In *Accident Analysis and Prevention*, pp. 109-216.
5. National Statistical Office, 1996, *Future Population Estimation*, Seoul (in Korean).
6. National Statistical Office, 2001, *Annual report on economically active population 2000*, (in Korean).
7. Simonds, R.H., and Shafai-Sahrai, Y., 1977, Factors apparently affecting injury frequency in eleven matched pairs of companies. In *Journal of Safety Research*, **9**, pp. 120-127.
8. United Nation, 1998, *World Population Prospects*.

Author Index

Agnew, J. 163
Akashi, M. 291
Amagai, K. 281
Arai, S. 89
Aridome, K. 129

Chisaka, H. 129
Cox, T. 119
Curbow, B.A. 163

Dote, T. 291

Edén, L. 99
Ejlertsson, G. 99

Fröhner, K-D. 137
Fukaya, T. 65

Gee, G.C. 163
Goedhard, W.J.A. 9
Greller, M.M. 153
Griffiths, A. 111

Hachisuka, K. 129
Hasegawa, T. 271
Hisamune, S. 281
Hopsu, L. 213
Horie, S. 37

Iijima, T. 89
Ilmarinen, J. 21

Johansson, I. 143

Kawakami, M. 77, 89, 223, 253
Kawasaki, J. 281
Kim, B. 261
Kim, H-K. 65
Kim, J.H. 49
Kimura, N. 281

Kishida, K. 281
Klemola, S. 185, 213
Kobayashi, E. 65
Koizumi, C. 291
Kono, K. 291
Kumashiro, M. 1, 253, 271, 301

Laflamme, D.J. 163
Lee, K.S. 49, 261
Lee, S. 205
Leppänen, A. 185, 213
Lim, H-K. 301
Lin, Y-C. 245
Louhevaara, V. 185

Matsushima, Y. 129
McDonnell, K.A. 163
Mikami, K. 233, 253
Muto, T. 65

Nagata, H. 205
Näsman, O. 177

Ohashi, N. 193

Petersson, J. 99

Saeki, S. 129
Sakamoto, S. 89
Shibata, H. 65
Shibuya, M. 253
Shimizu, H. 291
Shimodaira, Y. 193
Stroh, L.K. 153
Sugihara, Y. 65
Sugisawa, H. 65

Takemura, J. 129
Toba, Y. 77, 89
Tominaga, M. 291

Tsutsui, T. 37

Usuda, K. 291

Vercruyssen, M. 55

Wang, E.M-Y. 245
Wang, M-J.J. 245
Wang, S. 171
Watanabe, M. 77
Watanabe, T. 291

Yamashita, A. 77, 89

Keyword Index

A

Accident characteristics 301
Accident frequency 301
Accident severity 301
Activities of daily living 55
Aerobic ability 96
Age 111, 185
Aged worker 40, 291
Agility 55
Aging 40, 65, 89, 129, 175, 193,
   213, 233, 253, 281, 291, 301
Aging workers 163, 223, 245
Agriculture 193
Anthropometric data 245

B

Balance 55

C

Care labour 205
Co-operation 143

D

Disability pension 99

E

Early retirement 99
Elderly 49, 261
Elderly care worker 205
Elderly worker 40, 65, 253
Elderly workers 55
Employment 49
Engineers and innovators 137
Epidemic 55
ERGOMA approach 280
Ergoma strategy 2
Ergonomics 55, 77
Exercise 96
Exercise composition 77
Exercise intensity 77

F

Fishery work 281
Fitness 55, 185
Flexibility 55
Fluoride 291
Functional autonomy 55

G

Geriatrics 55
Gerontechnology 55

H

Health 55, 111, 185
Health maintenance 77
Health promotion 65
Health status 65
Healthy companies 233
Healthy life expectancy 9
Hemodialysis 291
Human aging 55
Human factors 55
Human resources 2

I

Inactivity 55
Income 49
Independent living 55
Information technology 223

J

Japan 65
Job 49
Job enlargement 223
Job enrichment 223
Job redesign 271

K

KAIZEN 233
Kitchen work 214
Know-how database 233

L

Late career 153
Legislation 41
Life expectancy 55
Lifespan functional fitness 55
Lifestyle 65
Lipid 96

M

Maintaining work ability 214
Male worker 96
Management 111
Middle-aged 40
Mobile phone 261
Musculoskeletal disorders 99

N

New work system 223
Nursing homes 205

O

Obesity 55
Occupational accidents 301
Occupational hazard 175
Occupational health 65
Older worker 41, 153
Older workers 119, 147

P

Policy 41
Post-polio syndrome 129
Power 55
Prescriptive exercise 77
Productive aging 2
Productivity 223
Products standard 193
Professional qualification 214
Promotion 185
Psychosocial factors 111

Q

Quality of life 99

R

Research and development 137
Retirement 55
Risk management 119

S

Shortage of engineers 137
Small-to-medium sized factory 280
Strength 55
Stress 111

T

Technology 55
Telecommunications 163
Terminology 37
Total productivity 177

U

Usability 261

W

Well-being 177, 214
Wellness 55
Work 171
Work - family balance 153
Work ability 2, 9, 55, 177, 185
Work accidents 281
Work burden 253
Work capability 193
Work characteristics 214
Work climate 163
Work situation evaluation 137
Work-related stress 119
Worker health 119
Working life expectancy 9
Workload 223
Workplace 129
Workplace policies 163

Y

Younger workers 147

Indices

9 780367 454692